IEE MONOGRAPH SERIES 7

POWER DIODE AND THYRISTOR CIRCUITS

D1826265

REX M. DAVIS

POWER DIODE AND THYRISTOR CIRCUITS

Published in association with
THE INSTITUTION OF ELECTRICAL ENGINEERS

Published by Peter Peregrinus Ltd.,
Southgate House, Stevenage, Herts. SG1 1HQ, England

First published in casebound edition 1971
© 1971 : Institution of Electrical Engineers

Republished in paperback edition, with minor corrections, 1976
© 1976 : Institution of Electrical Engineers

Reprinted, with minor corrections and additional References, 1979
©1979 : Institution of Electrical Engineers

ISBN: 0 901223 90 5

Typeset at the University Printing House, Cambridge
Printed in Great Britain by offset lithography by
Billing & Sons Ltd, Guildford, London and Worcester

CONTENTS

v

FOREWORD

The monograph aims at meeting two specific needs: that of the student and that of the practising engineer who wishes to become familiar with this rapidly expanding subject.

For the student, the author has combined three years' experience of teaching power electronics at the final-year undergraduate and graduate levels at Nottingham University. The book is ideally suited for courses at these levels at universities and technical colleges, since it provides a complete coverage of the subject in moderate depth with 135 references to the more advanced or the most recent developments. Students have found the 'voltage–time area' approach to the subject readily understandable, grasping the principles underlying circuit operation which have hitherto been considered complex. The extensive use of line diagrams illustrating the waveforms occurring in circuits enables the student to write down by inspection the mean output voltage of a rectifier circuit, for example.

For the engineer, the author has drawn on ten years of industrial experience in the design of thyristor equipment, from low-power applications such as a.v.r.s to variable-frequency invertors rated at up to 500 kVA. While the book is primarily concerned with the principles of static power convertors and controllers, the treatment is essentially practical, relating circuit design, operation and protection to the capabilities and limitations of the devices. In particular, the likely areas of interaction between power and control circuits are described in Chapter 9.

Devices with nonlinear (rectifying) characteristics have been available since the beginning of electrical engineering. The first device to prove suitable for power applications was the mercury-arc rectifier. The theory of rectifier engineering developed around the mercury-arc rectifier, with or without grid control, and usually multianode, which was cheaper than several single-anode rectifiers because a common cathode, with its ignition and arc-maintaining auxiliaries, served several anodes. This encouraged the development of circuits suited to the multianode rectifier, i.e. the star or centre-tap configuration. The books of that day (Marti and Winograd, 1930; Rissik, 1935) emphasise this arrangement, and remain important works of reference for rectifier engineers.

The silicon p–n junction diode, and later the thyristor, have revolutionised rectifier engineering in two ways. First, both diodes and thyristors are essentially single-anode devices requiring no ignition or excitation equipment, and requiring, in the case of the thyristor, only a very low-power firing pulse; the bias in favour of centre-tap circuits has disappeared, giving the circuit designer a new freedom. Secondly, the switching behaviour, though similar to that of the mercury-arc grid-controlled rectifier, is typically an order of magnitude faster, allowing hitherto uneconomical techniques to be exploited: these techniques are particularly centred around forced turn-off in d.c. circuits, using capacitors.

Although initially expensive, the solid-state devices have now reached price levels at which uncontrolled rectification is one of the cheapest of electrical-engineering processes, at about £0·50/kW; and controlled rectification, though more expensive than uncontrolled, is substantially cheaper than the motor–generator set alternative. Furthermore, solid-state devices with current ratings of 1–500 A are now available, allowing a far wider power spectrum to be covered economically using the same few basic circuit configurations.

The consequent upsurge of interest in rectification, inversion, regulation and power control using solid-state devices provides the reason for the publication of this monograph.

1979 reprinting

The opportunity has been taken of incorporating a few minor corrections and a supplement to the bibliography, covering the years 1970 to 1979.

ACKNOWLEDGMENTS

The author wishes to thank his wife Margaret, who undertook the difficult task of typing a seemingly meaningless text. It is the author's hope that the reader will find the book somewhat more meaningful.

Thanks are also due to colleagues at Nottingham University for patience during 1969 towards the author's inevitable preoccupation with the writing of this book.

LIST OF SYMBOLS AND ABBREVIATIONS

α	firing-angle delay (measured from where a diode would begin to conduct)
α_e	exponential rate of decay of damped oscillations
α_μ	conduction delay caused by overlap
$\beta_1\beta_2$	common-emitter current gains
β	firing angle of advance ($180° - \alpha$)
δi_1	finite change of current i_1
ζ	damping ratio (of damped oscillations)
μ	overlap angle
μ'	overlap angle (pseudo-overlap)
ϕ	magnetic flux
ψ	turnoff angle (resonant turnoff circuit)
ω	angular frequency ($2\pi f$)
ω_0	natural angular frequency (of resonant circuit)
$\boldsymbol{A}$	general matrix
$\left.\begin{array}{ccc} a & b & c \\ a' & b' & c' \end{array}\right\}$	transformer-secondary phase windings
B	flux density
C	capacitance
e	source voltage (time-varying)
E	amplitude of sinusoidal source ($e = E\sin\omega t$)
f	frequency, Hz
f_r	ripple frequency
f_s	supply frequency
H	inductance (large value, e.g. load inductance)
h	inductance (smaller value, e.g. leakage)
$\boldsymbol{I}$	unit matrix
I	current
i	current (time-varying)
i.p.t.	interphase transformer

k	general integer, fraction or factor
l	inductance (usually resonant or to control di/dt)
L	general symbol for load (see list of subscripts)
M	m.m.f.
m	residual m.m.f.
m	general fraction
N	number of turns on winding
n	general integer, fraction or factor
P	power
p	pulse number
q	number of secondaries feeding rectifier
q	general fraction
R	resistance
r, s, t	primary or supply phase
s	Laplace variable
T	time (fixed value), period, repetition period
T	symbol for transformer (in Figures)
t	time (variable)
t_{off}	turnoff time
t_{RV}	reverse-voltage time
t_1, t_2 etc.	successive instants of time
V	voltage
v	voltage (time-varying)
VA	apparent power
VAr	reactive power
VTA	voltage–time area
W	energy
x	percentage transformer reactance
y	general fraction
1–9	thyristors (in Figures and text)
11, 12	thyristors (auxiliary for turnoff only) (in Figures and text)
1'–8'	diodes (in Figures and text)
9'	freewheeling diode (in Figures and text)
2–3–2	conduction pattern in which two, then three, then two devices conduct together

EXAMPLES OF USE OF SUBSCRIPTS

e_a	voltage of phase a
I_{b2}	base current of transistor 2
I_{c1}	collector current of transistor 1
V_{C1}	voltage across capacitor at t_1, or initially
$I_{2'}$	current in diode 2'
μ_1	overlap angle between two upper (odd-numbered) devices in a bridge
μ_2	overlap angle between two lower (even-numbered) devices in a bridge
$\mu_{1'2}$	overlap involving diode 1' and thyristor 2
$\alpha_1 (\alpha_2)$	firing-angle delay for upper (lower) thyristor in a bridge
V_g	gate voltage
I_{hB}	increase of magnetising current, or buildup current, in h
$VTA_{i.p.t.}$	voltage–time area supported by interphase transformer
I_k	leakage current
V_L	voltage across load
ϕ_m	residual flux generated by residual m.m.f. m
i_m	magnetising current
I_d	mean output current
I_{dn}	minimum I_d
V_{d0}	mean output voltage at no load
I_{dt}	transitional I_d
I_{Rp}	peak I_R
I_{rr}	r.m.s. of I_r
VA_{rst}	total apparent power for phases r, s and t
V_s	supply voltage
I_{dsc}	I_d at its short-circuit value
I_{dx}	maximum I_d
V_1	voltage across thyristor 1
V_{14}	voltage at bridge terminal fed by thyristors 1 and 4

V_{L35} load voltage during interval from t_3 to t_5

VTA_{h12} voltage–time area across h between t_1 and t_2

h_{14} inductor associated with thyristors 1 and 4

V_{d1} output voltage in mode 1

I_{d12} output current at the boundary of modes 1 and 2

di/dt rate of rise of current, frequently in a thyristor at turnon; the primed symbol i' is also used for di/dt

dV/dt rate of rise of voltage, frequently across a thyristor at turnoff, or later

Plate 1. Selection of power diodes, thyristors and triacs, with associated heat sinks for natural-convection air cooling, forced air cooling and liquid cooling

A brief description of special features and electrical ratings which are commercially available at time of writing is given in Table 1 (by permission of AEI Semiconductors Ltd., ASEA (GB) Ltd, International Rectifier (GB) Ltd. and Westinghouse Brake & Signal Co. Ltd.) The cards with capital letters are 3·5 cm wide

1 INTRODUCTION: SEMICONDUCTOR DEVICES

1.1 Semiconductor materials

The outstanding properties of modern semiconductor devices arise from the characteristics of the junction between p type and n type semiconductor material within a single crystal (Gentry *et al.* 1964). Although the following consideration of current-carrier behaviour is of necessity superficial, it does provide some basis for the observed behaviour of semiconductor devices. In a pure silicon (or germanium) crystal, each atom with its own four outer electrons shares one outer electron with each of the four adjacent atoms. All outer electrons are thus occupied in covalent bonds between atoms. Only a very small number of thermally freed electrons (and vacated spaces or holes) are free to move under the action of an applied voltage to constitute a leakage current. Pure semiconductor material consequently has a very high resistivity and can be considered as an insulator.

Doping a pure semiconductor crystal with a trivalent or pentavalent impurity results in a very small proportion of crystal locations being occupied by atoms with three or five outer electrons instead of by silicon. In either case, the resistivity of the material is markedly reduced. For a pentavalent impurity, this is obvious, as free electrons are now readily available. The same is true for a trivalent impurity, since a hole is easily filled by an electron from an adjacent silicon atom, and so on, so that the hole effectively moves in the opposite direction to a sequence of hole-filling electron movements. The doped material is still electrostatically uncharged, as the charge of the current carriers introduced by the impurity atoms is cancelled by the opposite charge on their nuclei.

A junction is obtained by diffusing or alloying a higher concentration of one impurity partly into a wafer of silicon uniformly but lightly doped with the opposite impurity. Once formed, an isolated junction supports two opposed, balanced conduction mechanisms. As the junction is approached, the hole density in the p type material is reduced to match the low hole density in the n type material: there

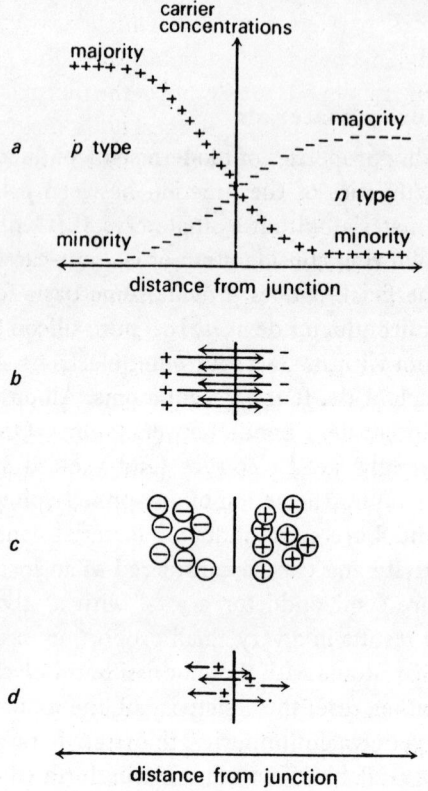

Fig. 1 Carrier concentrations and current-flow mechanisms in a p–n junction

a Carrier concentrations: + for holes, − for electrons
b Carrier movement resulting from concentration gradient
c Space charge on immobile impurity atoms developed by (b), restricting the above carrier movement
d Movement of thermally freed carriers induced by the space-charge barrier potential

is a diffusion of holes under the concentration gradient from the p type material across the junction to the n type material, where recombination takes place. The same is true for electrons traversing the junction in the opposite direction. Consequently, the regions near the junction do not have a sufficient carrier concentration to neutralise the fixed charges on the impurity atoms. In this region, the p type material acquires a negative, and the n type material a positive, space charge. A potential barrier now exists which restricts the diffusion process to the higher-energy carriers until the diffusion-current density equals the current density attributed to the thermally freed carriers in the junction region; these electrons and holes move in the opposite direction to those of the diffusion process, under the influence of the barrier potential. Fig. 1 illustrates the two mechanisms.

In fact, one cannot segregate electrons or holes into two groups—those generated thermally and those originating from higher-concentration regions where they are majority carriers—and claim that they behave differently. The crystal-lattice vibrations impart velocities to all electrons which are in most cases far greater than the drift movements described above. The actual movement of a particular electron is governed almost entirely by its energy derived from lattice vibrations. The behaviour in the junction region of electrons and holes, however generated, can nevertheless be divided into the two patterns described above.

1.2 Diode

The power diode consists of a p–n junction wafer with one surface soldered (or firmly pressed) to a substantial copper base (with, in the case of soldering, suitable intermediate materials to buffer differential expansion), and the other surface similarly attached to a counter electrode. The base forms one electrical connection and the path for heat to flow to an external cooling structure. The base also supports an enclosing insulator through which the other electrical connection passes to the counter electrode, and which hermetically seals the wafer from external contamination. Larger devices are now also contained in the disc package, which allows both sides of the wafer to be cooled. A selection of diodes, thyristors and cooling fins are shown in Plate 1 and described in Table 1.

The diode characteristics result from the modification of the

Table 1

Code (i)	Manufacturer (ii)	Construction Cooling (iii)	Maximum transverse voltage Mean current (180°) Stud temperature (except where stated) (iv)	Remarks and special features (v)
A	W	S AN	1200 V 200 A 120 °C	Diagonal fins allow vertical without derating. (a) Heat flows radially through device rim to diecast heat sink
B	W	IP AN	1500 V 500 A 65 °C	Diagonal fins as A, serrated to increase turbulence and surface area, cool flat-base thyristor with lug cathode
C	W	S AN	1200 V 200 A 120 °C	Aluminium fins cut from sheet material cool stud or flat-base devices
D	W	XP AF	700 V 550 A 85 °C	Plastics-encapsulated thyristor pressed between two heat sinks for double-sided cooling
E	W	XP AF	700 V 370 A 70 °C	As above, with single-sided cooling
F	IR	S	1200 V_{RRM} 200 A 125 °C*	Logic triac: single wafer passes and inhibits current in both directions * junction temperature
G	IR	S	2000 V_{RRM} 250–350 A 125 °C*	Stud-mounted thyristor * junction temperature
H	IR	XP	2000 V_{RRM} 470 A 125 °C*	Ceramic thyristor for double-sided cooling ('hockey puck') * junction temperature
I	AEI	IP AN	1500 V 140 A 85 °C	Circular flat-base thyristor, which can be rotated for best orientation of cathode lug (b) Rotatable anode lug avoids current in heat sink. (c) Co-axial gate connector and lead

Table 1 (*cont.*)

(i)	(ii)	(iii)	(iv)	(v)
J	W	XP	2500 V	Ceramic thyristor pressed between
			525 A	two heat sinks for double-sided
		AF	80 °C	cooling
K	ASEA		2200 V	Stud-mounted thyristor
			230 A	(*c*) Serrated fins increase turbu-
		AF	90 °C	lence and surface area
L	W	IP	4800/600 V	Flat-base diode
			425/840 A	
		AF	100 °C	
M	AEI	IP	2300/800 V	Flat-base thyristor
			200/400 A	
			85 °C	
N	W	IP	4800/2600 V	Circular flat-base thyristor
			425/550 A	(*e*) Anode is insulated from
			100 °C	heat sink by a beryllium cup
O	ASEA	XP	3500 V	Flat-base diode with insulation
			350 A	arranged radially, unlike
			105 °C	previous devices
P	ASEA	XP	3500 V	Clamping arrangements for O
			350 A	(*f*) Central screw for applying
		AF	105 °C	pressure to diode
Q	ASEA	XP	2200 V	Both sides of thyristor are cooled
			335 A	(see also K and O)
		AF	35 °C*	* air temperature
R	W	S	1200 V	Diodes mounted on two hollow
			200 A	liquid-cooled busbars
				System can be readily ex-
				tended for many parallel devices
		LF	35 °C*	* liquid temperature
S	AEI	XP	1500 V	Inverse-parallel thyristors forming
				a.c. switch
		LF	700 A	Thyristors are clamped between
			40 °C	liquid-cooled heat sinks

S = soldered; XP = external-pressure bond; IP = internal-pressure bond; AN = natural-convection air-cooled; AF = forced air cooled; LF = forced liquid cooled

5

balanced current-carrier mechanisms of an isolated junction by an externally applied voltage. A reverse voltage (p type negative) increases the barrier potential and hence suppresses the diffusion current, but leaves the thermal current largely unaltered. A virtually constant leakage current exists for a varying reverse voltage, until the electric stress in the crystal accelerates electrons so as to free others by collision. At this level, reverse leakage current rises rapidly, and avalanche breakdown occurs. For most diodes, the reverse-current density in the wafer is far from uniform, owing to lattice imperfections and varying depth of diffusion or alloying. Consequently, damaging local heating can occur, although the total power dissipation is small compared with the forward-current capability of the diode. Avalanche devices have a far more uniform reverse-current distribution, allowing steady reverse dissipation similar to the forward-current value, and transient reverse dissipation many times the steady value. Surface breakdown at the wafer's edge is becoming more important as device voltage ratings increase. Various bevelling techniques have been used to reduce the voltage gradient (Knott *et al.* 1965; Otsuka, 1969).

A forward bias (p type positive) reduces the barrier potential, thereby increasing the diffusion current. When the barrier potential is cancelled, the current rises rapidly as carriers cross the junction in both directions without hindrance. Typical diode characteristics are shown in Fig. 2 for cold and hot junctions.

An important phenomenon occurs when a conducting diode suddenly has a reverse voltage applied to it. The circuit inductance controls the rate of decay of current, and at zero current the junction is inevitably still flooded with carriers when the rate of decay of current is fast. A short period of reverse current is therefore possible, while these stored carriers reverse direction. They are rapidly swept out of the junction region, resulting in an abrupt interruption of reverse current. The circuit inductance will generate a large reverse voltage across the diode at this interruption, which can sometimes be enough to damage the diode (Blundell *et al.* 1961).

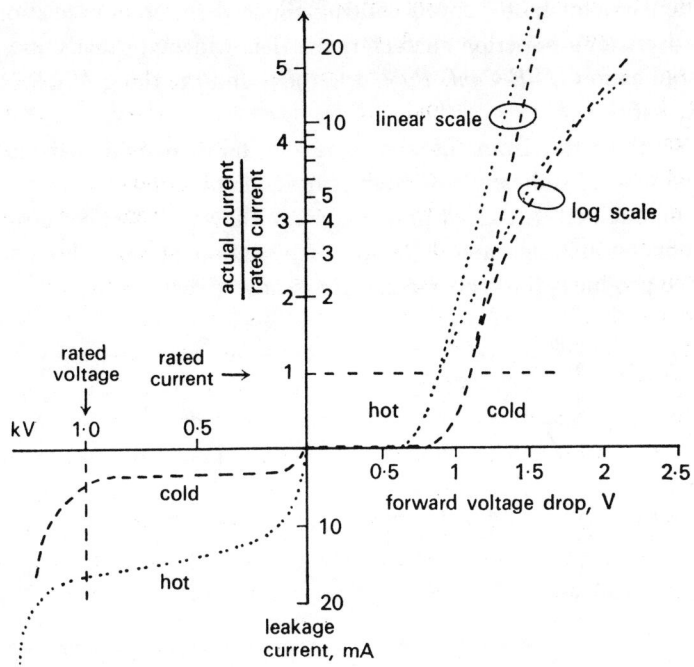

Fig. 2 Typical forward and reverse characteristics for a silicon diode
with junction temperatures of 20 °C and 100 °C

Note linear and log forward-current scales, and change of scales in reverse
quadrant

1.3 Thyristor

The thyristor is a 4-layer 3-junction device with two power
terminals and one for control. The mounting and encapsulation are
generally as for the diode.

Functionally, the thyristor differs from the diode in that forward
conduction is inhibited until a firing pulse is applied to the control
electrode called the gate. Subsequently, its behaviour resembles that
of the diode until forward conduction ceases, when it reverts to the
inhibited state. Thus the thyristor's behaviour closely resembles that
of the thyratron and mercury-arc grid-controlled rectifier, and, in
early applications, the thyristor was used as a replacement for these,

7

particularly for motor speed control. Since then, an appreciation of the thyristor's superior characteristics has widened greatly its field of application (*IEE Conf. Publ.* 17, 1965, and 53, 1969; *IEEE Conf. Rec.*, 1965).

The thyristor, from the above, is a bistable switching device: a short pulse is sufficient to switch it to its conducting state, which is self-maintained. A lack of power current allows the device to regain its nonconducting state. A simple explanation of how the 4-layer device produces these characteristics is now given.

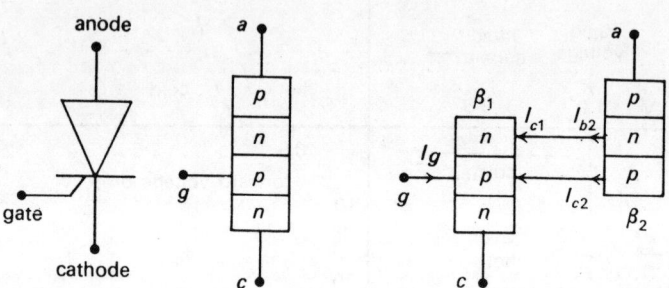

Fig. 3   Symbol and 2-transistor equivalent circuit for an element of the thyristor wafer

The explanation is based on the '2-transistor analogy', but let it be stated at the outset that, while this is a convenient user approach, it is an oversimplification. One of its shortcomings can be lessened by appreciating that a thyristor junction is of large area and does not possess uniformly distributed properties. In considering the 2-transistor analogy, it is necessary to consider a multitude of parallel 2-transistor analogies with differing properties, in order to explain some observed behaviour.

The 2-transistor analogy is illustrated in Fig. 3, which represents some elemental portion of the wafer. The transistors have common-emitter current gains β_1 and β_2, giving rise to the following relationships:

$$I_{c2} = \beta_2 I_{b2} = \beta_1 \beta_2 I_{b1}$$

Thus as I_g increases, β_1 and β_2 both increase until I_{c2} so adds to I_g to produce a regenerative turnon. As the anode current is reduced, β_1 and β_2 both decrease, so that, in the absence of a gate current

8

I_g, I_{c2} will provide insufficient current to maintain the regenerative effect and the thyristor turns off.

The position of the elementary transistor pair governing the latching and holding currents depends on circuit conditions. For fast turnon with rapidly rising anode current, the pair governing turnon are bound to be near the gate connection. At turnoff, the conduction area shrinks to the pair for which $\beta_1\beta_2$ is a maximum.

In practice, turnoff by gradual reduction of the anode current is rarely used. Applications generally reduce the anode current quickly, and the same carrier storage phenomenon as for the diode allows the anode junction momentarily to pass a reverse current. This junction is rapidly swept free of carriers, and a rapid interruption of reverse current occurs with the same possibility of generating a carrier-storage voltage spike. The circuit generally then establishes across the now nonconducting thyristor a reverse voltage which must be maintained for a period longer than the thyristor's turnoff time, which is typically 10–300 μs. During this period, the carriers trapped in the central junction region are too numerous for the leakage current of the anode junction, and recombination proceeds. The carrier concentration must be small enough when forward voltage is reapplied for the consequent device current, comprising leakage and displacement (junction capacitance), to maintain $\beta_1\beta_2$ below unity, otherwise the thyristor will refire, i.e. break over.

Thyristors are given a dV/dt rating of typically 10–200 V/μs, for the rate of reapplication of forward voltage after a prescribed reverse-voltage time not less than the turnoff time. Clearly, a longer reverse-voltage time allows the thyristor to sustain a somewhat greater dV/dt. For steady nonconducting conditions with a substantial forward anode voltage, short transients on this voltage with a high positive dV/dt may cause the thyristor to turn on, and, generally, the same dV/dt rating applies to this condition.

Particularly in forced-turnoff circuits using capacitors, thyristors can sustain very high rates of rise of current at turnon. As the regenerative turnon process propagates across the wafer slowly (Somos and Piccone, Cordingley, 1969) compared with the rate of rise of current for these applications, a very small portion of the junction

9

area near the gate is called on to carry the full device current (Mapham, 1963). Local heating and damage may result. Thyristors are therefore given a *di/dt* rating, typically 20–300 A/μs, to avoid this problem. Extra large gate signals with fast risetimes ($< 1\ \mu$s) greatly assist the thyristor's ability to handle a high *di/dt* safely. The problem of high *di/dt* frequently occurs when a thyristor takes over current from a conducting diode, where the thyristor and diode are together in series opposition across a low-impedance source. The insertion of series inductance is then a necessary remedy. More information is available from the handbooks [Gutzwiller (Ed.), 1967; Diebold (Ed.), 1962; Murray (Ed.), 1963].

1.4 Triac

The triac is a relatively new power device with current and voltage ratings similar to those of the thyristor. It can inhibit or pass current in either direction, and thus provides a single device to perform what previously required two reverse-parallel thyristors. A single gate is provided. For some devices, a positive gate pulse initiates current flow towards the gate end of the device (as for the thyristor); a negative gate pulse initiates conduction in either direction. For other devices, gate pulses of either polarity initiate conduction in either direction.

Several junction features not utilised in the thyristor are required to construct a triac (Gentry *et al.* 1964). It is not considered appropriate to introduce these here as triac applications are still few compared with those for diodes and thyristors.

The turnoff behaviour (Bergman and Knott, 1965) is worthy of mention. The application of a reverse voltage will not turn a triac off since the current carriers present will initiate reverse conduction. The triac current must be reduced below the holding current (which is usefully ten times greater than that of a thyristor) for a period to allow recombination of carriers. This is readily achieved for power-frequency operation with resistive loads.

For inductive loads, the current continues into the reverse half-cycle of voltage. When the current ceases, the device voltage jumps suddenly to a substantial value, instead of rising sinusoidally from zero as for the resistive load. The shortened period of current below the holding value, followed by a high dV/dt, make for turnoff difficulties with inductive loads. A series CR circuit across the triac is generally an adequate cure.

Triac applications will not be included separately in later Chapters (e.g. Section 6.4) since, in so many cases, the triac is merely an alternative device for a thyristor and diode, or for a reverse parallel-thyristor pair. Having a single gate, the triac clearly offers circuit simplifications, and immediate application areas are in a.c. power control. The dynamic properties are not, at the time of writing, sufficiently specified to allow its use in circuits requiring fast-turnoff devices. The cycloconvertor would clearly benefit from the replacement of thyristors by half the number of triacs. In general, the application of triacs does not introduce any new principles.

2 DEVICE APPLICATION PRACTICES

2.1 Cooling

Although the silicon wafer without electric stresses can withstand a temperature of several hundred degrees Celsius, leakage current becomes very pronounced above 150 °C, setting a practical limit to the working junction temperature. Diodes generally operate up to a junction temperature of about 200 °C; but controlled devices, because of their internal current gains, rarely retain their characteristics at a junction temperature above 150 °C. As the heat dissipation, caused almost entirely by forward-current and forward-voltage drop, greatly exceeds what the device can itself dissipate at these temperatures, the device must be mounted on a cooling fin or heat sink, as shown in Plate 1.

As the junction temperature is difficult to measure, a cooling design technique based on thermal resistance (deg C/W) has been developed. The heat generated at the junction is readily calculated from measured device waveforms, and each device type has a specified maximum thermal resistance between junction and base whereby the temperature rise between base and junction can be calculated. Extruded cooling-fin sections have their performance similarly specified in deg C/W for various lengths and for natural or forced air cooling, allowing the temperature difference between the device mounting position and the ambient air to be calculated. An allowance is also made for the contact thermal resistance between the base and fin.

The steady-state design above neglects the fact that devices are not conducting continuously, but are alternately passing pulses of current and sustaining reverse voltage. The thermal mass of the junction is very small, resulting in significant variations of temperature even with power-frequency operation. This has great application importance, as it highlights a fundamental difference between the semiconductor device and other power electrical equipment. The thermal time constant of a semiconductor device is at least an order of magnitude shorter. If junction-temperature variations occur within a 10 ms halfcycle, how much more will the junction temperature rise under what would normally be considered a short overload condition lasting 100 ms or so? Where semiconductors and machines are used together, an overload duty cycle on the motor is unlikely to require a larger motor power rating, but it will almost certainly require a larger rating for the semiconductor equipment.

Although the time constant of the semiconductor device is relatively short, it is not negligible. If the semiconductor equipment is only called on to support an overload from cold, or from a mean junction temperature well below the maximum, a useful overload capacity is available. Such capacity is utilised through the design concept of transient thermal resistance (Newsam, 1969). This specifies a lower thermal resistance for a short-duration surge of dissipation than for steady state, and incidentally indicates the extent of junction-temperature excursions in response to short pulses of dissipation.

While aluminium extrusion provides heat sinks for most applications, water- or oil-cooled busbars are used where the volume of cooling fins becomes excessive, e.g. for some large rectifiers for electrolytic applications. Manufacturers' handbooks [Gutzwiller (Ed.), 1967; Diebold (Ed.), 1962; Murray (Ed.), 1963] describe the practical aspects of mounting and cooling devices.

A recently developed device encapsulation (Warburton et al. 1966) allows the wafer to be cooled from both sides by pressing the device between two cooling fins, one for the anode and one for the cathode. This method offers the prospect of single devices with current ratings of more than 500 A. Pressure on the silicon wafer and its electrodes applied by the external cooling fins through the disc-type encapsulations is relied on for both electrical and thermal contact. Lack of soldering allows transverse relative motion to relieve differential expansion stresses. Pressure bonding is also used within most larger devices.

2.2 Series operation

In spite of the rising voltage ratings of thyristors and diodes, currently around 1·5–4 kV and 2–5 kV, applications still exist which require devices in series. The growing likelihood of thyristor application in d.c. transmission brings the possibility of long series strings.

The problems of series operation arise largely from the wide spread of device characteristics. Voltage-sharing techniques (Hey) will be described for thyristors, where there are many more factors to consider. In comparison, diode voltage sharing is simple. The spread of leakage current at breakover is such that, even for a steady string voltage, one thyristor can be on the point of breakdown while others are supporting only a small part of their rated voltage. Worst-case conditions are always assumed, i.e. one low-leakage device in series with $n-1$ high-leakage devices. The parallel sharing resistor for each device is given by

$$R = \frac{nV_T - V_S}{(n-1)(I_{kx} - I_{kn})}$$

where
$$V_T = \text{device voltage rating}$$
$$V_S = \text{string voltage}$$
$$I_{kx,\,kn} = \text{maximum, minimum leakage current.}$$

Selection allows the power loss in sharing resistors to be greatly reduced, but the possibility of device failure and the need for fast (unselected) replacement make it less attractive.

The sharing of transient voltage demands the use of a sharing capacitor in parallel with each device. The worst condition to consider occurs when the string is suddenly reverse-biased for turnoff purposes (Hall, 1969). There is a spread in the reverse-conduction times among thyristors of a type, and the first to cease passing reverse current will alone support the string reverse voltage. The parallel capacitor must carry the reverse current to the rest of the string until they also clear reverse current, without a dangerous rise of capacitor voltage. The allowable capacitor-voltage rise leaves a sufficient margin for the fast device to support, in addition, its equal share of the remaining prospective reverse voltage. The capacitor required, typically 0·1–1 μF, would, with thyristors, produce an excessive di/dt at turnon, so that some series resistance, typically 5–50 Ω, is also required: it is also important to minimise stray inductance in their connections. As the reverse carrier-storage currents can be appreciable, the voltage drop across the resistor must be added to the peak capacitor voltage when considering the allowable voltage across the

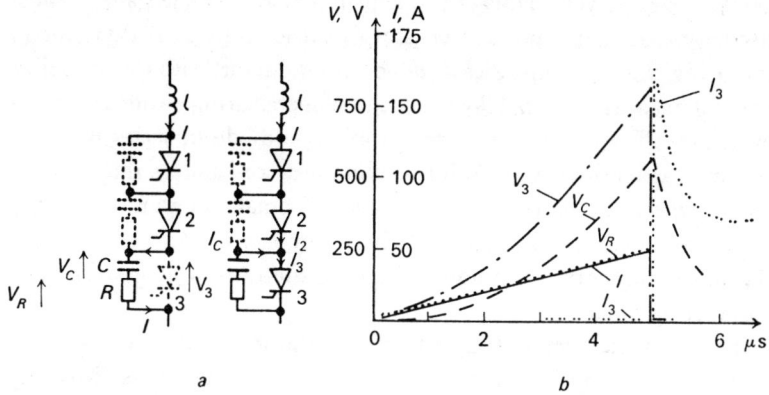

Fig. 4 Transient-voltage-sharing network (a) for a thyristor string, showing the high inrush current (b) to the slowest thyristor when it begins to conduct. The dotted parts of the circuits carry no current. A high inrush current of 175 A (for given conditions) occurs when the slow thyristor fires 5 μs later than the rest

$$R = 5 \,\Omega; \; C = 0·2 \,\mu F; \; dI/dt = 10 \, A/\mu s$$

fast device at the instant when the slow devices clear reverse current. With diodes, no series resistance is required.

A similar condition of device overvoltage can occur on firing a series string. A device which is slow to turn on in a string of fast devices will be left supporting the string voltage. Provided that the irregularity of turnon time is less than that of carrier storage time, the same capacitor network as above will suffice. The rise of voltage across the slow device and its capacitor is governed by the current, whose rate of rise must be limited to the capability of the thyristor. A single inductor for the string might be considered: this limitation will assist the control of a slow-device voltage, but it aggravates the di/dt turnon problem for the slow device. Fig. 4 shows an example

15

in which the series inductor limits di/dt to 10 A/μs, and the slow (bottom) thyristor 3 is assumed to turn on 5 μs late. The rise of current in the resistor and capacitor is shown as the solid line, while the voltages across R, C and 3 are shown dotted, dashed and chain-dotted, respectively. When 3 fires, it immediately accepts the 50 A flowing through the other thyristors; but, in addition, the discharge current of C through R and 3 of 125 A. An inrush current of 175 A would progressively damage the silicon wafer near the gate of most thyristors. Thus it may be necessary instead to fit each thyristor in the string with its own series di/dt control inductor, with the capacitor sharing network in parallel with both. Where sharing either at turnon or at turnoff is the predominant problem, a diode in series with the sharing capacitor may be helpful. Nonlinear resistances may also be useful where the nominal string voltage is small compared with the aggregate device voltage rating, when they can be designed to pass the full string current without exceeding the rated voltage of a non-conducting device.

To minimise turnon-time variations, the gate pulses must be as large as the ratings allow, with a risetime faster than 1 μs. Meeting this condition with a multisecondary pulse transformer for a long series string, with adequate insulation, can be difficult. Individual pulse transformers are generally preferred. Alternatively, synchronised firing using light-sensitive gate-firing devices provides a simple method for firing a long string, where the firing-pulse power is taken from the sharing capacitor.

Some thyristors can be reliably broken over or driven to limited reverse conduction without harmful effects (Gutzwiller; Yanai, 1965). This greatly simplifies the voltage-sharing arrangements, removing the necessity for both parallel resistors and capacitors.

2.3 Parallel operation

There appears to be a limit to the ultimate current rating of a single silicon wafer, both in cooling and in obtaining the required uniformity of characteristics throughout. The connection of devices in parallel to act as one will always remain a necessary technique.

While forced current sharing using series resistors or reactors to swamp differences in device forward-voltage drop is satisfactory, there are disadvantages, either of significant power loss, typically equal to that of the device, or of expense in the case of reactors.

Large-quantity production of high-current devices has led to the grading of forward-voltage drop, whereby devices of one grade share forward current to a specified tolerance of a few per cent., and clearly this is the preferred method. The sharing of surge currents does not present a problem, as the temperature coefficient of forward-voltage drop becomes positive somewhat above the rated current (Fig. 2).

The most significant problem occurs with parallel thyristors when the cluster current assumes low and then high values during the conduction interval. As in a single wafer, the conduction area shrinks at low currents, giving areas of wafer which are nonconducting; in a cluster, the voltage drop at low current across one thyristor may therefore be insufficient to keep another above its holding current. The difference between a wafer and a cluster occurs on the re-imposition of a large current: the wafer will naturally switch back into full conduction, but the nonconducting thyristors in a cluster cannot do so without gate drive. A gate signal lasting for the full conduction interval is therefore advisable for parallel thyristor operation. Depending on the rates of change of current expected, a prolonged train of pulses may be adequate. The layout of a cluster should be designed to equalise the total reactance of each path; otherwise the magnetic effect will tend to push the current outward to the paths in the weakest field.

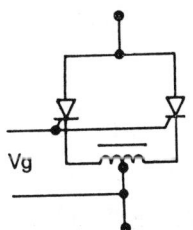

Where only two thyristors in parallel are required, the centre-tapped balancing reactor, connected as in Fig. 5, has the useful feature that, should one thyristor extinguish at a low current, a subsequent buildup of current in the other will generate a reactor voltage drop which, by auto-transformer action, will refire the extinguished thyristor (Dortort, 1958). Either one or two firing pulses can be used.

Fig. 5 Current-sharing reactor for two parallel thyristors

Where the circuit demands the limitation of di/dt on firing a cluster, it is helpful to provide each thyristor with its individual inductor, to force the sharing of di/dt, rather than allowing the fastest thyristor to bear the di/dt intended for the cluster.

2.4 Device failure mechanisms

Semiconductor devices require special consideration because of their low thermal mass, limited working junction temperature and nonuniform current distributions which occur at turnon and sometimes while blocking. The ways in which overvoltages and overcurrents occur in rectifier applications, and the appropriate protective techniques, are described in Section 5.6.

A well designed and correctly mounted device does not suffer from encapsulation or wafer-mounting failures either from base distortion or through thermal fatigue (a common source of failure with the earliest devices). Protection is thus reduced to the prevention of overheating caused either by a blocking failure due to local, bulk or surface breakdown, or by excessive forward current. The effects of overheating can be catastrophic, when immediate junction damage occurs, or progressive, when prolonged or repeated high temperatures increase the rate at which the contaminants inevitably present, particularly ions, migrate to the highly stressed junction edges where they cause a local deterioration of blocking properties.

2.4.1 Overvoltage breakdown

Reverse overvoltage causing local or surface breakdown requires only a small reverse power dissipation compared with the device forward rating to cause damage. When a device has bulk-avalanche properties, its steady-state reverse dissipation is similar to its forward rating, and, transiently, a much higher dissipation is tolerable, allowing such devices to absorb safely all but the highest-energy voltage transients.

Forward overvoltage will damage a thyristor if surface breakdown occurs before breakover. Such devices have a peak forward-voltage rating which, although greater than the minimum breakover voltage,

18

may be less than the actual breakover voltage of a particular thyristor. Surface contouring (Knott *et al.* 1965) is used to control the stress at the junction edges (Otsuka, 1969).

Protection against overvoltage breakdown is governed largely by the application. The only mechanism which generates an overvoltage solely related to the characteristics of the device is carrier storage. In most cases, the overvoltage protective arrangements described in Section 5.6, or voltage-sharing components where present, also cater for the carrier-storage voltage spike: otherwise, a capacitor, typically 0·1 μF, is required in parallel with the device, with, in the case of a thyristor, some series resistance, typically 5–50 Ω, to limit di/dt at turnon.

2.4.2 Overcurrent damage

Localised forward-overcurrent damage (Gentry, 1958) occurs when the rate of rise of current in a thyristor at turnon is excessive (Mapham, 1963 *a*) or when turnon by breakover occurs giving rise to a higher di/dt than the designer expected for gate turnon. For moderate current overloads, the whole device is progressively overheated, leading to excessive reverse leakage current and localised melting. A severe overload will cause immediate melting, destroying the junction. Repeated overloads which are tolerable singly can produce a degradation of characteristics. Accordingly, a peak repetitive current rating is given, below which degradation is not detectable. Infrequent surges above this current rating are allowable to simplify fuse protection (see also Section 5.6.3).

2.4.3 Fuselinks for semiconductors

Whereas overvoltage protective arrangements are usually particular to the circuit and application in which devices are used, overcurrent protection by fuselinks (Lerstrup, 1965; Jacks, Newbery, 1969) is a universal technique associated with the devices and is appropriately discussed here.

The power semiconductor device is given an I^2t rating which defines, in conjunction with its own conducting resistance, an allowable energy which will not damage the device on an infrequent basis. Conventional fuselinks, if chosen to suit the continuous (r.m.s.) rating of the device, will allow, under fault conditions, more than this critical energy to flow through the device before the fuse completes the interruption of fault current. Ranges of special fuselinks with a low ratio of I^2t to r.m.s. current rating have been developed, based on a notched silver-strip element. It is now possible to select a fuse-link with a lower I^2t rating than the thyristor, and yet having an r.m.s. rating which approaches closely—and sometimes exceeds—the r.m.s. current in the thyristor when it is carrying its rated mean current (Golden, 1968). To avoid significant device derating, it is usually necessary to place the fuselink in series with one device only, which precludes the use of an a.c. line fuse with a bridge rectifier.

Fuselinks for semiconductors also differ from their industrial counterparts in their arcing behaviour. As the peak-voltage rating of semiconductor devices is typically only twice the crest supply voltage, and occasionally less than this, it is important that the arc voltage of a fuselink clearing a faulty device in an equipment does not produce an overvoltage on other devices. Semiconductor fuselinks have a closely controlled arcing voltage, typically 1·2 times their peak-voltage rating. If the fuselink is being used in a circuit in which the voltage is appreciably less than the rated fuselink voltage, the actual I^2t will be less than the manufacturer's figure, but the arcing voltage will be greater than 1·2 times the circuit voltage.

Finally, it is important to distinguish between two fundamentally different philosophies of fuselink protection in equipment using few and many semiconductor devices. For small equipment, it is common to use the device fuselinks to protect the devices against external overloads or short circuits. For large equipment, the device fuselinks are present for one purpose alone, to isolate a faulty device and thus to allow the remainder to continue operation without interruption. Overcurrent and short-circuit protection is now provided externally either by a circuit breaker or by a fuselink. The equipment is designed so that the fuselinks of the devices do not operate on external over-loads or faults (Blundell *et al.* 1961).

3 BASIC RECTIFIER CIRCUITS: CENTRE-TAP RECTIFIERS

3.1 Introduction

The preface gives the historical background to the development of rectifier circuits. The essentially single-anode construction of the power semiconductor diode and thyristor (in contrast to the multi-anode mercury-arc rectifier) has removed from rectifier circuits any historical bias in favour of the centre-tap arrangement (Marti and Winograd, 1930; Rissik, 1935). Consequently, the bridge circuit, which minimises the rating of the transformer secondary, is favoured (Spreadbury, 1962; Schaeffer, 1965) except for low-voltage high-current supplies, where the smaller number of devices offsets the extra cost of the transformer (Read, 1945).

Throughout the book the circuit analysis neglects device forward-voltage drop and reverse leakage current. In Chapter 5, some reference to resistive loads is made, and some comparisons are drawn between rectifier-circuit behaviour on resistive and inductive loads; the analysis of Chapters 3 and 4 assumes a highly inductive load which provides perfect smoothing. An inductive load of quite modest time constant smooths the rectifier output current so that an assumption of perfect smoothing is acceptable. For higher powers, multi-phase rectifier circuits are used in which the component of ripple voltage is small and of a high frequency, validating further the assumption of perfect smoothing. Apart from that in the load, circuit resistances, including the resistance of transformer windings, are neglected in the analysis. Circuit inductance, including leakage inductance, plays an important role in circuit behaviour, and consequently the fundamental relationship of power-convertor circuits analysis is

$$di/dt = e/H \tag{3.1}$$

or, over a period T,

$$\Delta I = \int_0^T \frac{edt}{H} \tag{3.2}$$

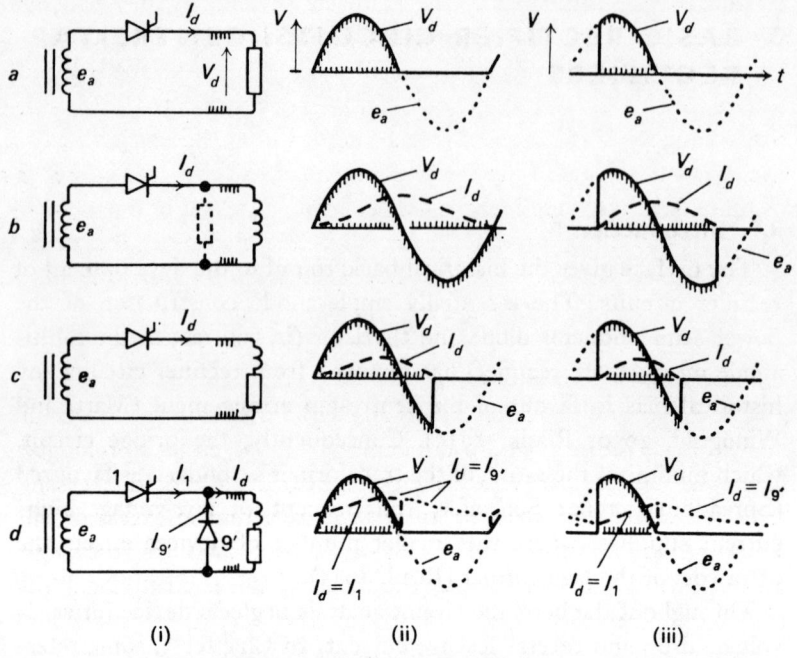

Fig. 6   Halfwave-rectifier behaviour

a   Resistive load
b   Purely inductive load
c   Inductive load with series resistance
d   As (c) with freewheeling diode
  (i)   Circuit arrangement
 (ii)   Output voltage and current for diode rectifier
(iii)   Output voltage and current for thyristor rectifier

where $\int_0^T e\,dt$ is the voltage–time area supported by the circuit inductance H during the period T.

Halfwave rectifiers are not used for power applications, but they
do serve as a useful introduction to device behaviour in simple circuits. The left-hand set of waveforms of Fig. 6 apply to a diode, and
serve to compare the output voltage V_d (solid) and current I_d (dashed)
for resistive and inductive loads. With a pure inductance (Fig. 6b),
the peak of I_d after the first positive halfcycle of voltage will be given

22

by the voltage–time area of the halfcycle divided by the load inductance. The inductance maintains current flow into and throughout the negative halfcycle, at the end of which the current is reduced to zero. For the more realistic case (Fig. 6c) of an inductive load with series resistance, the current reaches zero during the negative halfcycle, not at its end. The freewheeling diode shown in Fig. 6d is commonly used to take over the load current at the end of the positive halfcycle, and hence prevent the negative halfcycle from reducing the load current to zero. For long-time-constant loads the load current is continuous, supplied from the transformer during the positive halfcycles, and carried by the freewheeling diode during the negative halfcycles.

The right-hand set of waveforms apply to a thyristor fired after an angular delay α. For the pure inductance (Fig. 6b), either a prolonged gate pulse or some parallel resistance, shown dotted, would be essential for a pulse-fired thyristor, to provide a current in excess of the latching current. In Fig. 6d, the thyristor only conducts from α to π, and hence, for continuous current, the freewheeling diode will conduct from π to $2\pi + \alpha$.

Power-rectifier applications almost universally use sinusoidal supplies, either single or multiphase. Some basic voltage–time areas for various portions of a half sine wave are given in Fig. 7 in terms of E, the peak line/neutral or line/centre-tap voltage, and angular frequency ω. Also, where there is a delay in the start of conduction, it is important to know the voltage–time area lost for the delay. This is also indicated in Fig. 7. It is helpful to memorise the voltage–time areas shown in Fig. 7, as they are used regularly in the calculation of output voltages. In 3-phase circuits, the area subtracted for phase delay α is bounded above and below by phase voltages; consequently, the amplitude to be used is $\sqrt{3}E$, not E. This occurs first in eqn. 3.9.

The calculation of the mean output voltage for the many various rectifier circuits is based on finding the net output voltage–time area in a repetition period and dividing this by the period. In every case, controlled rectification is assumed. For uncontrolled rectifiers, the delay α is set to zero.

The general centre-tap arrangement is shown in Fig. 8. For $q = 1$, the circuit is halfwave and of little application importance. When q is even, the secondaries form centre-tap pairs, giving fullwave rectification of each phase. When q is odd, a polyphase halfwave rectifier results. The circuits for which $q = 2$, 3 and 6 are of most importance since, for higher values, the conduction angle of each device is progressively shortened, increasing the ratio of r.m.s. to mean device and

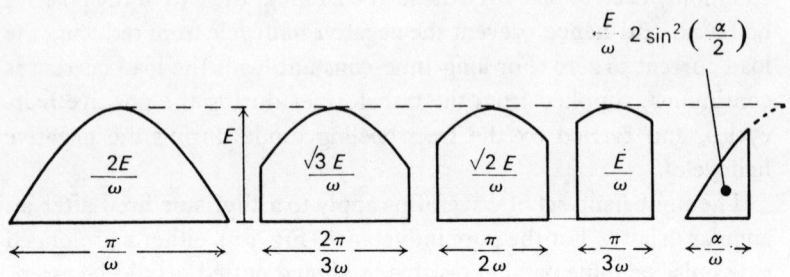

Fig. 7   Voltage–time areas for commonly occurring portions of a half sine wave

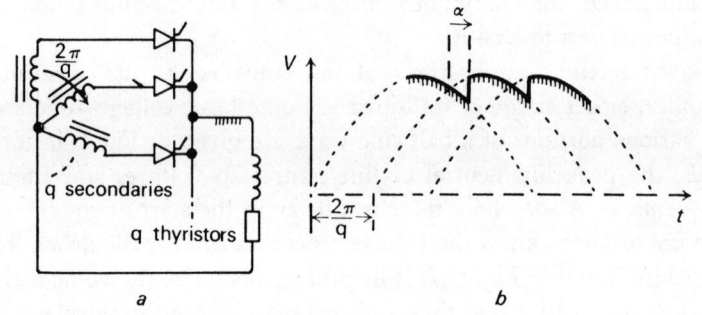

Fig. 8   Generalised centre-tap rectifier with q secondaries
a Circuit
b Output voltage for zero leakage inductance (no overlap) and delay angle α

secondary current and giving poorer utilisation. The value of q gives the ratio of the output ripple frequency to the a.c. supply frequency, a ratio commonly called the pulse number p.

The waveforms of Fig. 8 show how the output voltage V_d is generated for zero transformer leakage inductance, where the transfer of current from one device to the next is instantaneous.

24

Transformer leakage inductance slows down (Feinberg and Chen, 1964a, b) the transfer of current from one device to the next, causing a period of overlap when both devices conduct together. Fig. 9 shows the essentials. When only diode $1'$ conducts, I_d, assumed constant and smooth, produces no voltage drop in h_a, so that $V_d = e_a$. When e_b becomes positive with respect to e_a, a current $i_{2'}$ builds up, at the

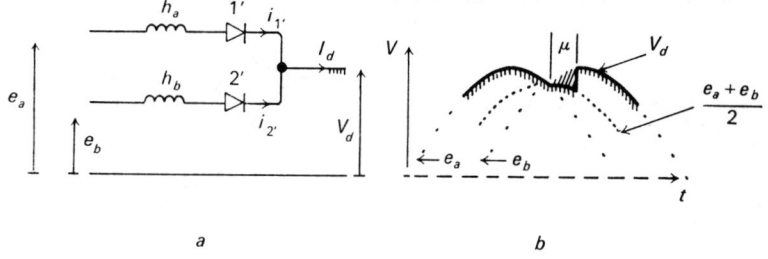

Fig. 9   Principle of overlap
a Two overlapping diodes
b Output-voltage waveforms before, during and after overlap μ
The output-voltage waveform V_d is boundary-shaded below the line

expense of $i_{1'}$ so that $i_{2'} + i_{1'} = I_d$. As the rates of change of $i_{2'}$ and $-i_{1'}$ are equal, the voltage drops across h_b and h_a are also equal, so that, during overlap, V_d is the mean value of e_a and e_b. Consequently, during overlap, the output-voltage waveform looses an area (shaded) necessary to change the current in h_a (or h_b) from its value at the start of overlap to that at its end. For Fig. 9, the lost voltage–time area (shaded) is

$$VTA_h = hI_d \qquad (3.3)$$

3.2   Centre-tap rectifiers: output parameters
3.2.1   Single-phase

The waveforms for a single-phase centre-tap circuit ($q = 2$) are shown in Fig. 10 for a smooth output current I_d and a total transformer leakage and supply inductance h, referred to each half secondary. Thyristors 1 and 2 are fired after a delay α from the position where diodes would naturally begin conduction. The assumption of continuous load current enforces the continued conduction of each thyristor into the region of reverse output voltage until the next thyristor is fired, which initiates overlap. During overlap, the output voltage is zero (the mean of e_a and e_b).

The voltage–time method of obtaining the output voltage is explained in detail for this example. The first term in brackets in eqn. 3.4, $2E/\omega$, is the area of a full halfcycle corresponding to no delay or overlap. The second term resulting from the firing delay is

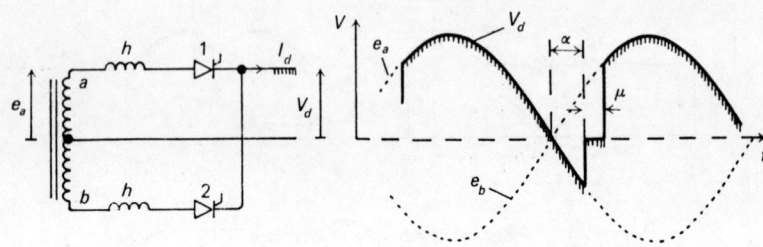

Fig. 10   The single-phase centre-tap ($q = 2$) controlled rectifier
The output-voltage waveform for delay angle α and overlap angle μ
is shown for smooth output current I_d

the triangular area of duration α, which is twice that in Fig. 7, since the prospective amplitude of the horizontally shaded area is $2E$. The third term is the area lost in overlap during the buildup of phase current from zero to I_d. The repetition period is π/ω. Eqn. 3.5 expresses the output V_d in terms of its no-load value and a voltage drop.

$$V_d = \frac{\omega}{\pi}\left(\frac{2E}{\omega} - 2\frac{E}{\omega}2\sin^2\frac{\alpha}{2} - I_d h\right) = \frac{2E}{\pi}\cos\alpha - \frac{I_d\omega h}{\pi}$$

$$(3.4), (3.5)$$

The overlap angle μ is given by

$$I_d h = \frac{E}{\omega}\left(2\sin^2\frac{\alpha+\mu}{2} - 2\sin^2\frac{\alpha}{2}\right) \qquad (3.6)$$

Sometimes it is desirable to connect a freewheeling diode in parallel with the rectifier output, as shown in Fig. 11. At the beginning of the negative halfcycle, as V_d cannot reverse, the reverse voltage behind the conducting thyristor decays the thyristor current, causing the diode current to build up at the same rate to maintain I_d constant.

26

This overlap between thyristor and diode will proceed until the thyristor current is zero (as shown), unless the next thyristor is fired first. In the latter case, as at any instant the rate of decay of current in the outgoing thyristor is equal to the rate of current buildup in the incoming thyristor, and the current has already dropped from I_d in the outgoing thyristor when the incoming thyristor is fired, the diode current must remain constant during the triple overlap, and the out-

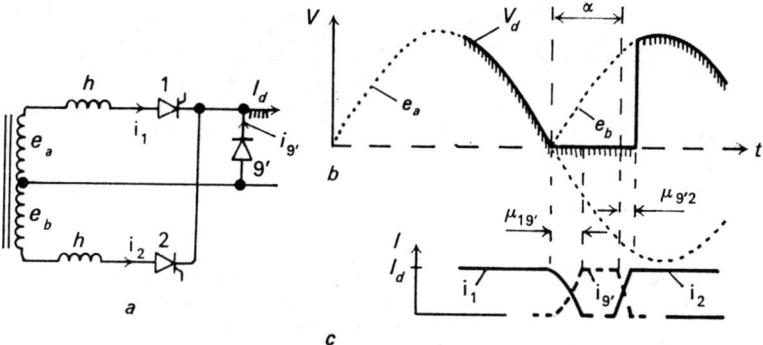

Fig. 11 Controlled rectifier ($q = 2$) with freewheeling diode
Note the absence of instantaneous reverse output voltage
a Circuit
b Output-voltage waveform V_d
c Device current waveforms

going thyristor ceases conducting first. This is followed by a simple overlap between the diode and incoming thyristor, as the current in the thyristor builds up at the expense of the diode current.

Regardless of the pattern of device conduction at overlap, the loss of output voltage–time area is the result of the firing delay α and area required to establish I_d in the incoming thyristor. The relationship is therefore

$$V_{d\alpha} = \frac{\omega}{\pi}\left(\frac{2E}{\omega} - \frac{E}{\omega}2\sin^2\frac{\alpha}{2} - I_d h\right) \tag{3.7}$$

$$= \frac{2E}{\pi}\left(\frac{1}{2} + \frac{\cos\alpha}{2}\right) - \frac{I_d\omega h}{\pi} \tag{3.8}$$

The term $(\frac{1}{2} + \frac{1}{2}\cos\alpha)$ in the second formulation is the same as that obtained for the half-controlled bridge rectifier (eqn. 4.5), described

27

in Chapter 4. The term indicates that a delay $\alpha = 180°$ is required
to reduce the output voltage to zero. A freewheeling diode always
increases the delay angle α required to reduce the output voltage
to zero, by preventing any reverse voltage–time area from occurring.
In the above case, the overlaps $\mu_{1\,9'}$ and $\mu_{9'2}$ are calculated from
eqn. 3.6 by substituting $\alpha = 0$ and $\alpha = \alpha$.

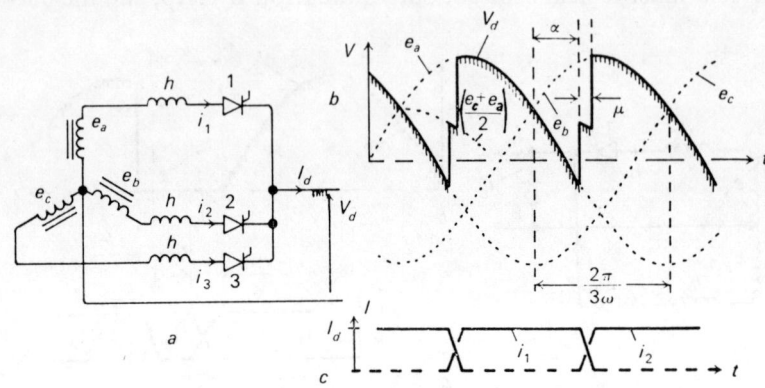

Fig. 12   The 3-phase centre-tap rectifier ($q = 3$)

a Circuit
b Output voltage
c Device current for delay angle α and overlap μ

3.2.2 3-phase

The 3-phase circuit ($q = 3$) is treated similariy. Fig. 12 defines
the terms and illustrates the output waveform of V_d. The output
voltage $V_{d\alpha}$ is given by inspection:

$$V_{d\alpha} = \frac{3\omega}{2\pi}\left(\frac{\sqrt{3}E}{\omega} - \frac{\sqrt{3}E}{\omega}2\sin^2\frac{\alpha}{2} - I_d h\right) \tag{3.9}$$

$$= \frac{3\sqrt{3}E}{2\pi}\cos\alpha - \frac{3I_d\omega h}{2\pi} \tag{3.10}$$

The voltage–time area lost by the delay α results from a line–line
voltage $\sqrt{3}E$, and the repetition period is $2\pi/3\omega$.

For angles of delay $\alpha < 120°$, the subtraction of the area corresponding to the delay from the no-delay area $\sqrt{3}E/\omega$ for a repetition

28

period is straightforward. For larger angles of α, the process is easier to visualise if it is noted that the two areas in Fig. 13 are equal. Alternatively, the waveform can be considered 'inverted', as shown in Fig. 14, which gives the output voltage in terms of the angle β as

$$V_{d\beta} = -\frac{3\omega}{2\pi}\left(\frac{\sqrt{3E}}{\omega} - \frac{\sqrt{3E}}{\omega} 2\sin^2\frac{\beta}{2} + I_d h\right) \qquad (3.11)$$

which reduces to the same expression, because $\beta = \pi - \alpha$. The details of invertor operation are given in Chapter 6.

Fig. 13　Equal voltage–time areas

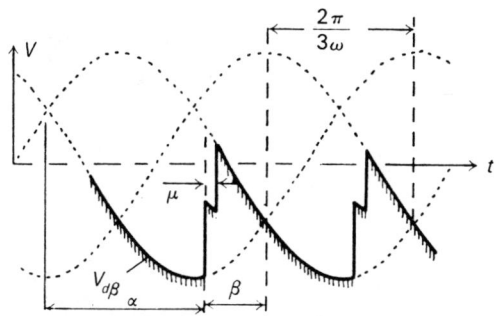

Fig. 14　Output voltage for centre-tap rectifier
$(q = 3)$ for $\alpha > 90°$

Inversion is indicated by the reversed mean output voltage. The angle of advance $\beta \ (= 180° - \alpha)$ is preferred for invertors

The effect of a freewheeling diode in parallel with the output is only felt for a delay angle $\alpha > 30°$, as is clear from Fig. 12. For a delay $\alpha < 30°$, the above relationship for $V_{d\alpha}$ holds. For $\alpha > 30°$ the first overlap occurs as the current in the outgoing thyristor transfers to the diode. If this overlap is complete before the incoming thyristor is fired, a second simple overlap takes place between the diode and this thyristor, and the waveform is as shown in Fig. 15.

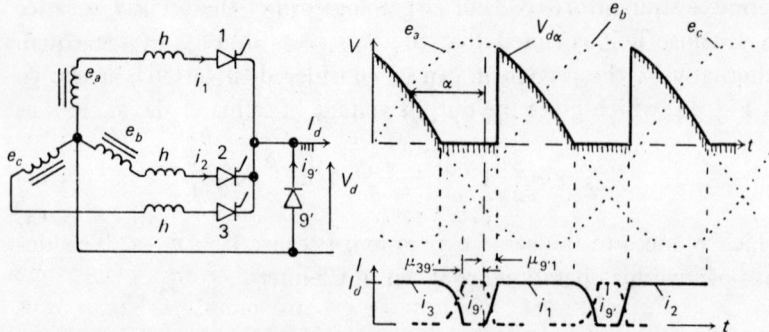

Fig. 15   Output voltage and device current for centre-tap rectifier
with freewheeling diode ($q = 3$)

Note the absence of instantaneous reverse output voltage

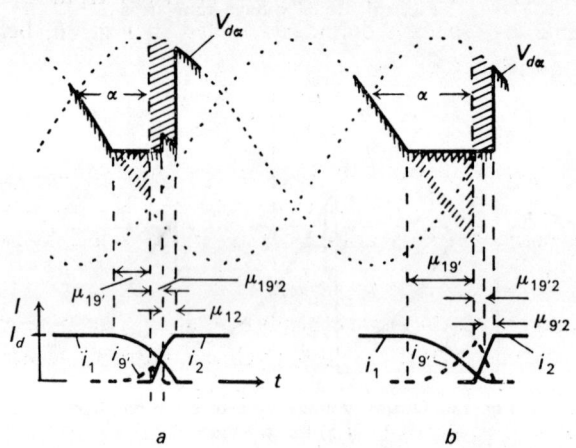

Fig. 16   Output voltage and device current for the
rectifier of Fig. 15, showing triple overlap

a Freewheeling diode ceases conduction first
b Outgoing thyristor ceases conduction first
In both cases, the shaded area is $I_d h$, and the boundary shaded area determines
the peak diode current

If, as in Fig. 16, the incoming thyristor is fired while the first
overlap is in progress, a triple overlap occurs, at the end of which
either the diode (Fig. 16a) or the outgoing thyristor (Fig. 16b) will
cease conducting first, depending on α. In either case, however, the

voltage–time area (shaded) lost after firing the thyristor is $I_d h$, since this is the area required to establish I_d in the incoming inductance h. The output voltage $V_{d\alpha}$ is therefore given by the relationship

$$V_{d\alpha} \atop \alpha > 30° = \frac{3\omega}{2\pi}\left(\frac{2E}{\omega} - \frac{E}{\omega}2\sin^2\frac{\alpha+30}{2} - I_d h\right) \qquad (3.12)$$

$$= \frac{3E}{\pi}\{\tfrac{1}{2} + \tfrac{1}{2}\cos(\alpha+30)\} - \frac{3I_d\omega h}{2\pi} \qquad (3.13)$$

This expression gives the same output voltage as the previous one for $\alpha = 30°$, but higher for $\alpha > 30°$. Consequently, a larger α is required to reduce the output voltage to zero, in this case 150°.

The calculations associated with the triple overlap must account for the two possibilities illustrated in Fig. 16. The diode current $i_{9'\alpha}$ at the firing instant α is readily calculated from the reverse voltage–time area, shown boundary-shaded in Fig. 16, behind the outgoing inductance:

$$i_{9'\alpha} = \frac{1}{h}\frac{E}{\omega}2\sin^2\frac{\alpha-30}{2} \qquad (3.14)$$

During the period of triple overlap, each supply voltage e_a, e_b produces a voltage–time area which causes a change of current δi_1 and δi_2 while the diode current varies from its initial value, to satisfy the relationship

$$i_{9'} = I_d - i_1 - i_2 \qquad (3.15)$$

Two expressions for the angle of triple overlap $\mu_{19'2}$ must be evaluated, one for each possibility; the lesser value is correct. Assuming that the diode current reaches zero first, we get

$$0 = i_{9'\alpha} - \delta i_2 - \delta i_1 \quad (\delta i_1 < 0) \qquad (3.16)$$

$$= i_{9'\alpha} - \frac{1}{h}\frac{E}{\omega}\left(2\sin^2\frac{30+\alpha+\mu_{19'2}}{2} - 2\sin^2\frac{30+\alpha}{2}\right)$$

$$+ \frac{1}{h}\frac{E}{\omega}\left(2\sin^2\frac{\alpha-30+\mu_{19'2}}{2} - 2\sin^2\frac{\alpha-30}{2}\right) \qquad (3.17)$$

Assuming that the outgoing thyristor current reaches zero first, we get

$$0 = I_d - i_{9'\alpha} - \delta i_1 \qquad (3.18)$$

$$= I_d - i_{9'\alpha} - \frac{1}{L}\frac{E}{\omega}\left(2\sin^2\frac{\alpha-30+\mu_{19'2}}{2} - 2\sin^2\frac{\alpha-30}{2}\right) \qquad (3.19)$$

Once $\mu_{19'2}$ has been calculated, i_2 and i_1 or $i_{9'}$ can be calculated at the end of the triple overlap. The remaining simple overlap $\mu_{9'2}$ or μ_{12} is calculated from the voltage–time area required to raise the current in the incoming inductance from i_2 to I_d. The two cases must be considered separately, since, when the overlap is between the out-going and incoming thyristor, the voltage–time area extends down to the mean of the two phase voltages e_a and e_b, provided that $(e_a + e_b)/2 > 0$; whereas, if it is between the incoming thyristor and the diode, or when $(e_a + e_b)/2 < 0$, the area extends down to the zero voltage level. These differing voltages are apparent from Figs. 16a and b. The corresponding relationship is

either $\quad I_d - i_2 = \dfrac{1}{h} \dfrac{\sqrt{3}E}{2\omega} \left(2\sin^2 \dfrac{\alpha + \mu_{19'2} + \mu_{12}}{2} - 2\sin^2 \dfrac{\alpha + \mu_{19'2}}{2} \right)$ $\quad$ (3.20)

or $\quad I_d - i_2 = \dfrac{1}{h} \dfrac{E}{\omega} \left(2\sin^2 \dfrac{30 + \alpha + \mu_{19'2} + \mu_{9'2}}{2} - 2\sin^2 \dfrac{30 + \alpha + \mu_{19'2}}{2} \right)$

$$(3.21)$$

The above example has been considered in detail to show how the necessary expressions can be written down by inspection, once the output-voltage waveform has been established, using only the simple rules of Fig. 7.

3.2.3 Summary

The dependence of no-load output voltage on firing-angle delay, i.e. the control characteristics, for centre-tap rectifiers is conveniently summarised in Fig. 17. The solid curve applies to the simple centre-tap circuit of any number of phases and illustrates the reversal of output voltages for delay $\alpha > 90°$. This reversal of voltage for continuous current indicates a reversal of power flow and the onset of inversion for which load energy is transferred back to the a.c. supply.

The characteristics for rectifiers with freewheeling diodes are shown dotted. For $q = 3$ and $q = 6$, the dotted curves depart from the solid at $\alpha = 30°$ and $\alpha = 60°$, since, for angles less than this, the output voltage is never negative and hence the diode can have no effect.

Where rectifiers are used to supply resistive loads, no reversal of output voltage is possible. The output-voltage waveform, and hence the control characteristic, are therefore similar to those obtained for a rectifier with a freewheeling diode which also prevents reversal of output voltage.

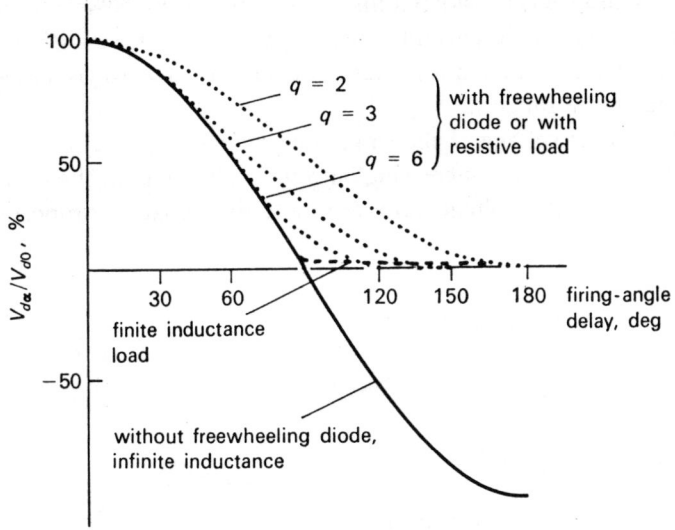

Fig. 17   Control characteristics for centre-tap rectifiers with loads of infinite and finite inductance, with and without freewheeling diode

An inductive load of finite time constant cannot maintain current flow indefinitely if the rectifier output voltage is zero or negative. The steady-state control characteristic for such a load is shown by the dashed curve which departs from the solid curve, when the output current becomes discontinuous, and its actual position depends on the load time constant.

The operation of rectifier circuits for other than perfectly smooth output currents is treated more fully in Chapter 5.

3.3 Centre-tap rectifiers: characteristics to short circuit

An excessive output current coupled with high leakage inductance substantially lowers the output voltage of a rectifier by prolonging the overlaps, giving poor voltage regulation. Rectifier behaviour for these conditions is considered for diodes only, as closed-loop current control or electronic current limiting prevents such conditions developing with controlled rectifiers. The output current I_d is, as usual, assumed to be smooth.

The output voltage of the single-phase circuit ($q = 2$) is reduced linearly to zero as the increasing output current extends the overlaps to 180° since the voltage–time area lost in overlap is proportional

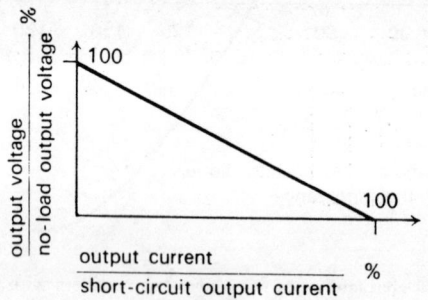

Fig. 18   Regulation characteristics to short circuit
for centre-tap rectifier ($q = 2$)

to the output current. The regulation characteristic is a straight line from the no-load voltage to the short-circuit current, as shown in Fig. 18. The shapes of the secondary phase-current waveforms for various overlap angles are shown in Fig. 19, the current waveform ending with a $1 - \cos\omega t$ waveshape for a short-circuit load.

With polyphase rectifiers ($q \geqslant 3$) the overlap between two diodes cannot extend to 180°, without the overlap voltage $(e_a + e_b)/2$ first becoming negative with respect to the next phase voltage e_c. When this occurs, a triple overlap develops during which the overlap voltage becomes $(e_a + e_b + e_c)/3$.

34

Regardless of the number of diodes conducting together or their sequence, the voltage–time area lost during the buildup of phase current $I_{1'p}$ is $I_{1'p}h$, where $I_{1'p}$ is the peak phase current. For small values of overlap, $I_{1'p} = I_d$; but, when $I_{1'p}$ fails to reach I_d, owing to extended overlap and conduction in one or more adjacent phases, the voltage–time area lost during the phase-current buildup becomes less than $I_d h$ in the ratio $I_{1'p}/I_d$. Consequently, the slope of the voltage-regulation curve is reduced.

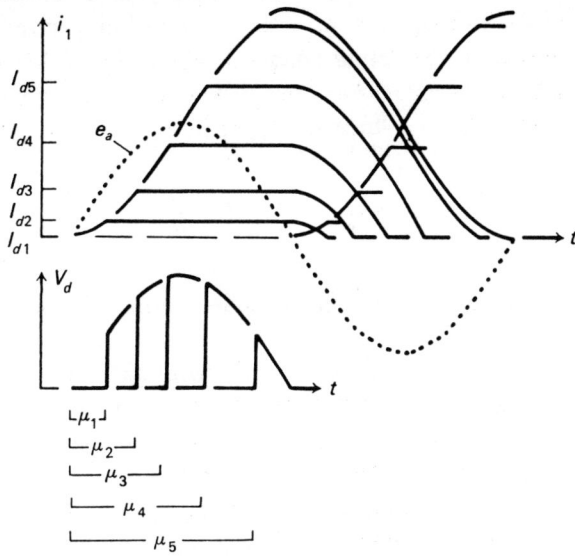

Fig. 19 Diode current and output voltage for progressively increasing overlap angle μ required by progressive rise of output current I_d

Whenever the minimum number of diodes conducting together increases by one, the slope of the regulation curve will be reduced. The reduced regulation can also be considered as due to the effective paralleling of leakage reactances by overlapping diodes. Load behaviour is categorised into modes, for which the mode number is the minimum number of diodes conducting together.

It should be noted that the overlap-angle definition used hitherto, namely the angle over which two devices share the output current by conducting simultaneously, becomes inadequate for multiple overlaps. To be consistent with the voltage–time area approach, the definition of overlap to be applied for multiple overlaps is taken as the angle between the start of phase-current decay at light load (i.e. the phase-voltage crossover) and the diode extinction. This definition has been used when indicating the overlaps μ in Fig. 20.

For the 3-phase case ($q = 3$), the minimum number of diodes conducting together changes from one to two for the condition illustrated in Fig. 20c, for which the overlap μ is 104.5°. For values of I_d giving an overlap $\mu \leqslant 90°$, the relationship derived earlier (eqn. 3.9) applies to the output voltage V_{d1} in mode 1:

$$V_{d1} \atop \mu \leqslant 90 = \frac{3\omega}{2\pi}\left(\frac{\sqrt{3}E}{\omega} - I_d h\right) \tag{3.22}$$

$$= \frac{3\omega}{2\pi}\left(\frac{\sqrt{3}E}{\omega} - \frac{\sqrt{3}E}{2\omega}2\sin^2\frac{\mu}{2}\right) \tag{3.23}$$

While eqn. 3.22 applies throughout mode 1, eqn. 3.23 becomes invalid for $\mu > 90°$, although mode-2 operation has not yet been reached, because the occurrence of a triple overlap alters the output-voltage waveform to zero ($e_a + e_b + e_c = 0$) during the triple overlap. Thus the equation describing the horizontally shaded area in Fig. 20a in terms of the overlap μ does not apply to the corresponding area in Fig. 20b, although the error is very small.

For angles of overlap between 90° (Fig. 20a) and 104.5°(Fig. 20c), the output voltage V_{d1} obtained from Fig. 20b is

$$V_{d1} \atop 90 \leqslant \mu \leqslant 104.5 = \frac{3\omega}{2\pi}\left(\frac{\sqrt{3}E}{\omega} - I_d h\right) \tag{3.24}$$

$$= \frac{3\omega}{2\pi}\left\{\frac{\sqrt{3}E}{\omega} - \frac{\sqrt{3}E}{2\omega} - \frac{E}{\omega}\left(2\sin^2\frac{\mu+30}{2} - 2\sin^2\frac{90+30}{2} + 2\sin^2\frac{\mu-90}{2}\right)\right\} \tag{3.25}$$

This gives the same output voltage as eqn. 3.23 for $\mu = 90°$.

For mode 2, shown in Fig. 20d, corresponding to angles of overlap $\mu > 104.5°$ there are always at least two diodes conducting together, so that the peak phase current $i_{1'p}$ no longer reaches the output

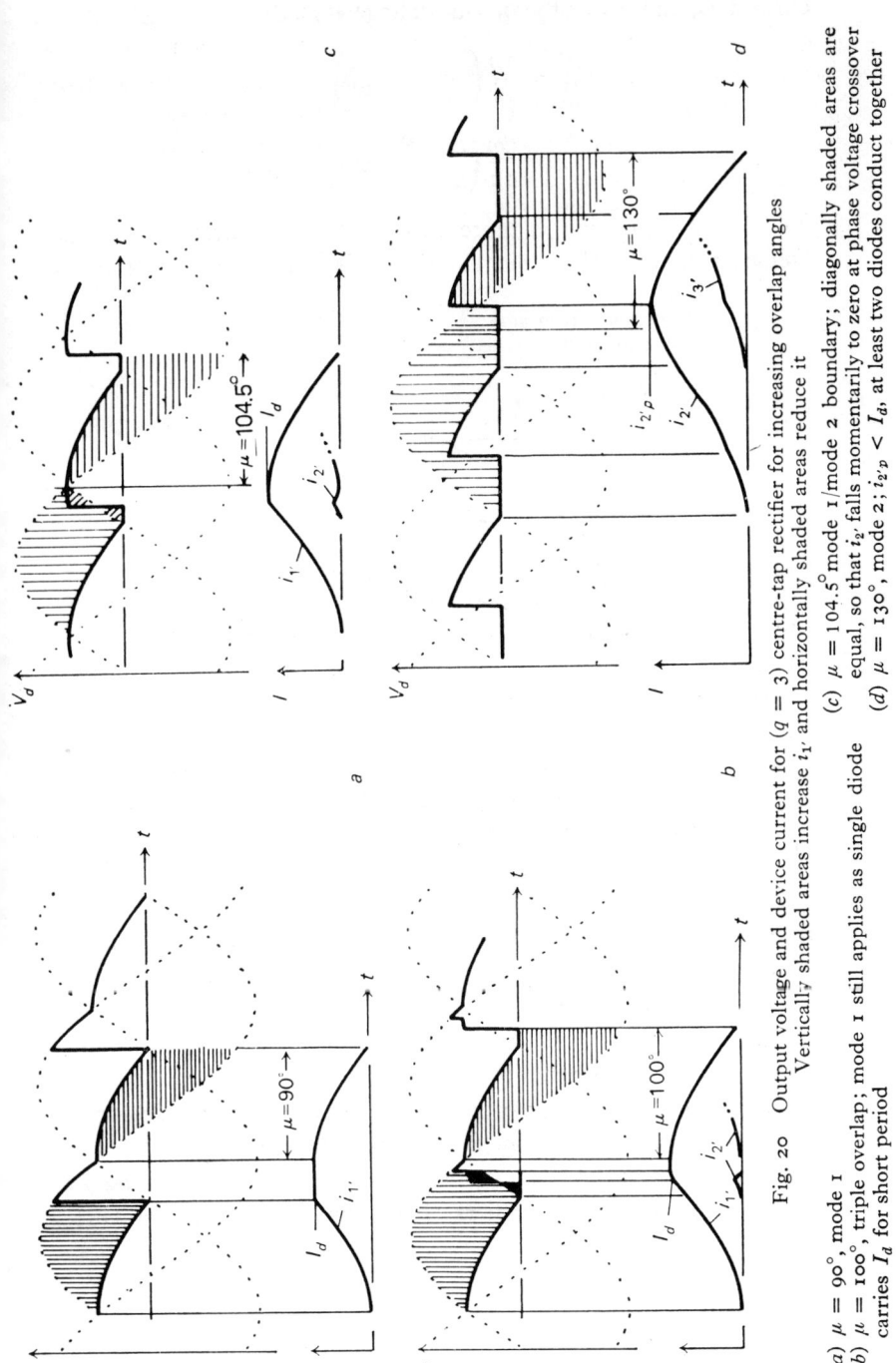

Fig. 20 Output voltage and device current for $(q = 3)$ centre-tap rectifier for increasing overlap angles
Vertically shaded areas increase $i_{1'}$ and horizontally shaded areas reduce it

(a) $\mu = 90°$, mode 1
(b) $\mu = 100°$, triple overlap; mode 1 still applies as single diode carries I_d for short period

(c) $\mu = 104.5°$ mode 1/mode 2 boundary; diagonally shaded areas are equal, so that $i_{2'}$ falls momentarily to zero at phase voltage crossover
(d) $\mu = 130°$, mode 2; $i_{2'p} < I_d$, at least two diodes conduct together

37

current I_d, and hence the output voltage V_{d2} will be

$$V_{d2}_{\mu>104.5} = \frac{3\omega}{2\pi}\left(\frac{\sqrt{3}E}{\omega} - i_{1'p}h\right) \qquad (3.26)$$

$$= \frac{3\omega}{2\pi}\left(\frac{E}{\omega} - \frac{E}{2\omega}2\sin^2\frac{\mu-30°}{2}\right) \qquad (3.27)$$

The latter expression is more easily calculated directly from the voltage–time area of output voltage. It is clear from Fig. 20d and

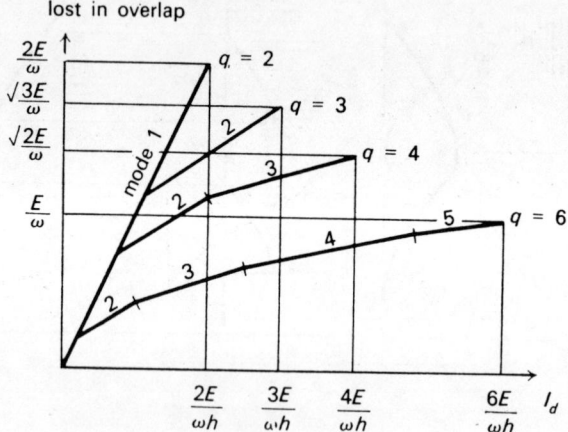

Fig. 21 Voltage–time areas lost in overlap for centre-tap rectifiers with $q = 2$, 3, 4 and 6 for all modes of operation

Characteristics terminate when the lost voltage–time area = voltage–time area at no load, and the output current is at its short-circuit value

from eqn. 3.27 that zero output voltage occurs when the overlap μ (as defined) reaches 210°.

For rectifiers where $q = 4$ and $q = 6$, there are three and five modes, respectively. As the voltage–time area lost in mode 1 is always $I_d h$ for any number of phases, it is helpful to compare graphically the voltage–time areas lost per repetition period, for various values of q, as shown in Fig. 21. The initial slope for mode 1 is the same for all values of q, but it becomes less for the higher modes of operation of polyphase rectifiers. The short-circuit condition corre-

38

sponds to a lost voltage–time area equal to the no-load voltage–time area, and the short-circuit current for any value of q is given by

$$I_{dsc} = \frac{qE}{\omega h} \qquad (3.28)$$

The calculation of the overlap angle and output voltage at a mode boundary follows the method of Fig. 20c. The voltage waveforms which apply at the mode boundaries for a $q = 6$ rectifier are illus-

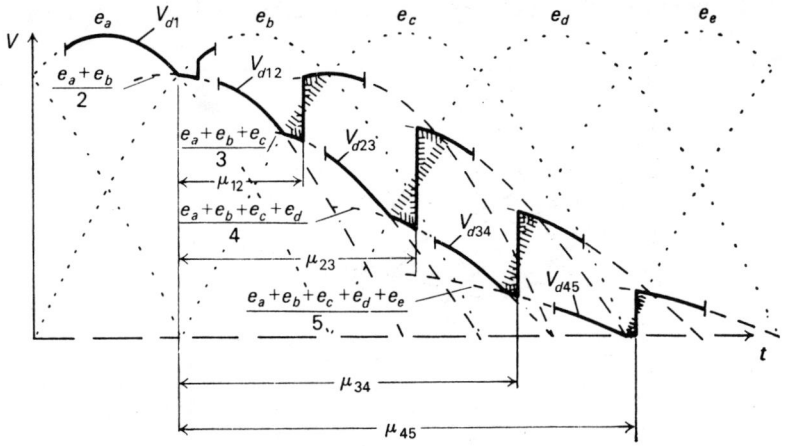

Fig. 22   Output voltage of a rectifier ($q = 6$)
at the mode boundaries

Only one repetition period is shown in each case. V_{d23} indicates the output voltage on the boundary of modes 2 and 3

trated in Fig. 22, each bold waveform being shown for one repetition period only. The left-hand waveform is mode-1 operation and is included for comparison. The four remaining waveforms are for boundary conditions between modes 1 and 2, 2 and 3, 3 and 4 and 4 and 5. The criterion for a mode boundary is the equality of area of the two shaded triangular regions. In the lower shaded region, current will build up in an incoming phase, but it will be collapsed to zero in the upper shaded region, as previously illustrated for the mode-1/ mode-2 boundary in Fig. 20c.

39

The formation of the normal voltage-regulation characteristics from the above is now straightforward, and is expressed nondimensionally in Fig. 23 for centre-tap rectifiers having $q = 2$, 3, 4 and 6.

3.4 Rectifier transformers and apparent-power (VA) ratings

The rectifier transformer is an essential part of a centre-tap rectifier circuit. To calculate the load placed on the a.c. supply by a centre-tap rectifier and its transformer, the behaviour of the transformer and the way in which the primary-current waveforms (Wallach, 1963) are

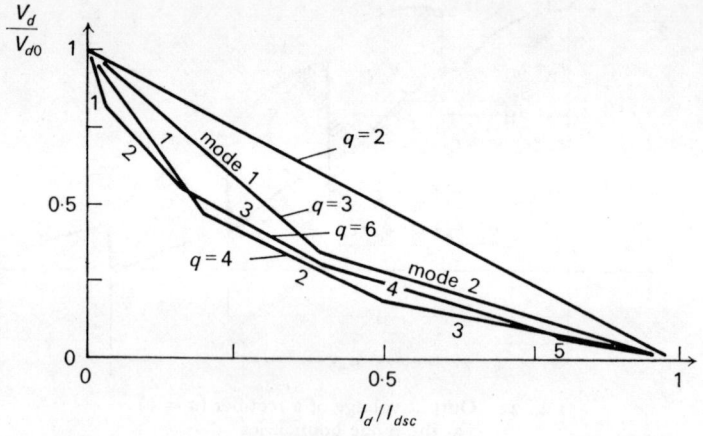

Fig. 23 Voltage-regulation characteristics for centre-tap rectifiers with $q = 2$, 3, 4 and 6, between no load and short circuit, showing $q - 1$ modes

obtained from the known output current I_d must be appreciated. In the previous Section, it has been assumed that the e.m.f.s behind the secondary leakage inductances are sinusoidal, which is not true for certain polyphase-transformer connections. This Section examines rectifier transformers for centre-tap circuits, and shows how primary currents and apparent-power ratings of windings may be obtained.

3.4.1 Winding currents and m.m.f. conditions

Single-phase and 3-phase transformers are illustrated in Fig. 24, in which the core permeability is assumed to be infinite. The m.m.f.s M, M_a, M_b, M_c produced by the combined winding currents on the transformer legs obviously take the values shown. With the simple 3-phase core, any residual m.m.f. m sets up a flux ϕ_m through the core, which returns largely outside the primaries and secondaries, frequently via the tank. Inphase voltages are generated in the windings according to $N(d\phi_m/dt)$. These voltages are in addition to those dropped across leakage inductances.

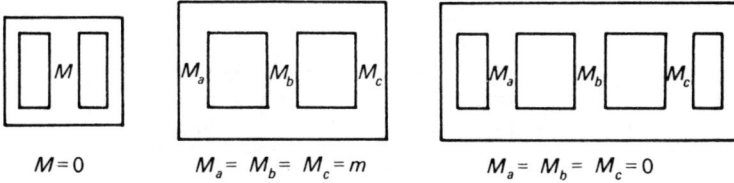

$M = 0$ $\qquad M_a = M_b = M_c = m$ $\qquad M_a = M_b = M_c = 0$

Fig. 24 Transformer cores used frequently in rectifier transformers, showing m.m.f.s per leg and the possibility of a residual m.m.f. m in the 3-phase 3-leg core

To introduce the concepts which can arise in rectifier transformers, a very simple case will be considered first. Fig. 25 shows a halfwave rectifier ($q = 1$) feeding an inductive load with a freewheeling diode. (This simple circuit finds application in the supply of contactor coils.) Since the supply voltage has no d.c. component, and the induced e.m.f. in the primary can have no d.c. component in the steady state, it is clear that there can be no d.c. component in the steady-state primary current. Assuming infinite load inductance and zero transformer leakage inductance, the secondary current i_a will be as shown in the Figure. For a $1:1$ ratio, this will also represent the load current reflected into the primary windings. This current clearly has a d.c. component $I_d/2$, and hence the magnetising current must have an equal and opposite d.c. component, so that the total primary current has none. A magnetising current with a d.c. component is only possible if the B/H loop of the core material is used asymmetrically as shown. The greater the load current I_d, the further into saturation the flux excursions must go to provide the equal and opposite d.c. component. It is immediately apparent that the two halves of the B/H loop cannot be used equally, otherwise the d.c. component would be zero; likewise, the magnetising-current waveform must be asymmetrical, as shown in the Figure, resulting in the primary-current waveform indicated.

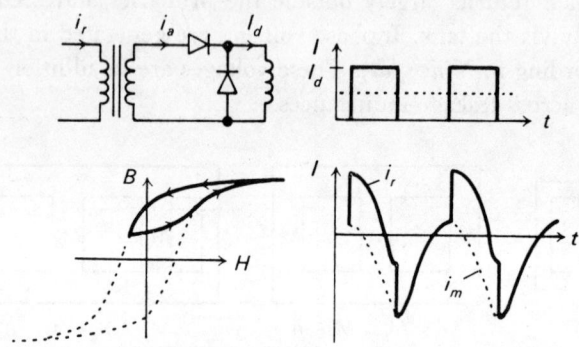

Fig. 25 Asymmetrical use of core material imposed by a secondary
current which contains a d.c. component

Magnetising current i_m consequently has an equal
and opposite d.c. component

For single-phase rectifiers where $q = 2$, the secondary m.m.f. alternates, resulting in an alternating primary m.m.f. in opposition. Perfect cancellation of load m.m.f.s allows the core to behave symmetrically about the centre of the B/H loop, giving full utilisation of the core.

The understanding of polyphase transformers is more difficult, owing to the variety of flux paths and winding connections. For a star primary, intermittent secondary currents can produce a residual m.m.f. which, by inducing inphase voltages in all windings, modifies the waveforms and the diode conduction pattern. For a delta primary, the inphase voltages, and the residual m.m.f. producing them, are suppressed by a circulating current in the delta, which invalidates the usual factor $\sqrt{3}$ relating phase and line currents as the triplen harmonics are missing from the line currents. It is the avoidance of these phenomena which has led to multiple rectifier circuits described in Chapter 5.

The above features are well illustrated in the diametrical 6-phase-rectifier transformer ($q = 6$) with a star primary, in which all windings are assumed to have the same number of turns N. Fig. 26 illustrates the circuit, the secondary currents, the residual m.m.f., the induced inphase voltage e_m and the line current i_r. The residual m.m.f. m is given by

$$m = M_a = M_b = M_c \qquad (3.29)$$

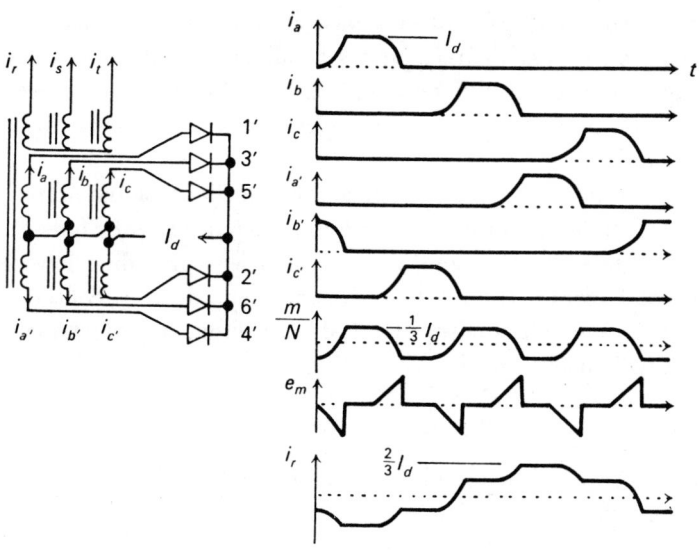

Fig. 26 Diametrical 6-phase centre-tap rectifier ($q = 6$) showing secondary-current waveforms, residual current ($= m/N$), the triplen harmonic voltage e_m and the primary current i_r

Since
$$m/N = i_r + i_a - i_{a'} = i_s + i_b - i_{b'} = i_t + i_c - i_{c'} \qquad (3.30)$$

and
$$i_r + i_s + i_t = 0 \qquad (3.31)$$

$$3m/N = i_a - i_{a'} + i_b - i_{b'} + i_c - i_{c'} \qquad (3.32)$$

The last relationship establishes the waveform and amplitude of the residual m.m.f., from which the waveform of the line current i_r can be obtained with eqn. 3.30.

The inphase triplen harmonic voltage e_m is shown for the upper secondaries with respect to the starpoint; the lower secondaries have a reversed polarity. By examining the overlap between diodes 1′ and 2′, this induced voltage will be seen to increase e_a and reduce e_c, which will prolong the overlap and steepen the voltage-regulation characteristic. When mode-2 operation is reached, there are always two diodes conducting, and the decay of current in the outgoing diode 1′ is brought about by the rise of voltage behind diode 3′. As the induced harmonic voltage is in phase in the phases a and b (Fig. 26), it does not influence the overlap, so that the slope of the regulation curve becomes less, for two reasons. First, the total change of phase current i_{ap} is less than I_d; and secondly, the voltage e_m does not influence the voltage difference $e_b - e_a$. It can be said that the presence of e_m hastens the onset of mode-2 operation with its flatter regulation characteristic.

If the voltage e_m can be obtained for a smaller residual m.m.f., the onset of mode 2 will occur at a lower output current. This can be achieved using a 5-leg core, the two outer legs providing low reluctance paths for the residual flux necessary to produce e_m, consequently requiring a much smaller m.m.f. In the limit, a perfect magnetic shunt will ensure that $m = 0$, so that

$$0 = i_a - i_{a'} + i_b - i_{b'} + i_c - i_c \tag{3.33}$$

which is only satisfied if a minimum of two diodes conduct together, i.e. in mode-2 operation.

The presence of a triplen frequency residual flux ϕ_m can cause undesirable heating in the tank, core and clamps, so that a circuit which has no residual m.m.f. is to be preferred.

Consider next the star–star ($q = 3$) rectifier. The diodes now conduct for at least $120°$ giving the waveform shown in Fig. 27. The m.m.f. relationship yields

$$m/N = i_r + i_a = i_s + i_b = i_t + i_c \tag{3.34}$$

Since
$$i_r + i_s + i_t = 0 \tag{3.35}$$

$$3m/N = i_a + i_b + i_c \tag{3.36}$$

44

The last relationship, with the waveforms of i_a, i_b and i_c, indicates that the residual m.m.f. has an amplitude and waveform as shown producing a steady flux. There is thus no induced e.m.f. e_m, but a steady magnetisation of the core. According to the reluctance of the flux path for ϕ_m, some core saturation will occur, reducing core utilisation and increasing losses.

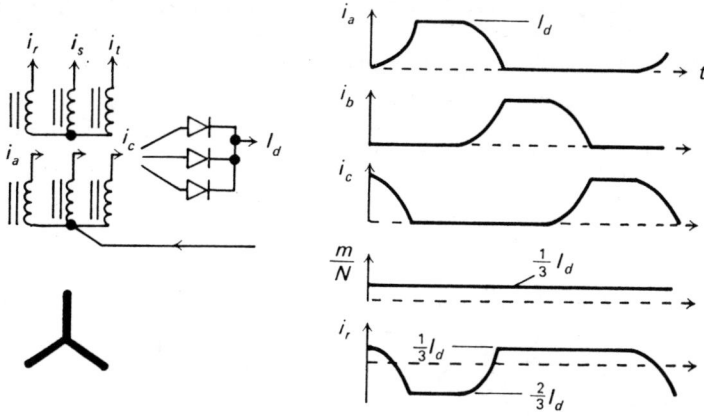

Fig. 27   3-phase centre-tap rectifier ($q = 3$) showing secondary-current waveforms, a constant residual current m/N and the primary current i_r having half-cycles of unequal amplitude

A 5-leg core in this case would force m to be zero, so that each primary and secondary current would have identical cancelling waveforms—both with d.c. components. An asymmetrical magnetising current would result with a d.c. component equal and opposite to that of the reflected secondary current, giving marked saturation, poor core utilisation and increased losses.

Two rectifier circuits with $q = 3$, having separate secondaries on the same core, can be series- or parallel-connected to provide cancellation of their residual m.m.f.s. One form of series connection in which the secondaries become common results in the 3-phase bridge circuit, which is always used in preference to the other form.

A third form involves the series connection of secondary windings only, as shown in Fig. 28, for which the diode-current waveforms

45

$i_{1'}$, $i_{2'}$ and $i_{3'}$ are the same as i_a, i_b and i_c of Fig. 27. Note that the lower half secondaries are wound in the reverse sense. Eqns. 3.37–3.39 below for residual m.m.f. indicate that this has been cancelled, but the price paid is a reduction of output voltage, owing to the vector addition of half-secondary voltages to $(\sqrt{3})/2$ of that obtained with single secondaries, increasing the secondary rating per unit output.

$$m/N = i_r+i_{1'}/2 -i_{2'}/2 = i_s+i_{2'}/2 -i_{3'}/2 = i_t+i_{3'}/2 -i_{1'}/2 \quad (3.37)$$

Since
$$i_r+i_s+i_t = 0 \quad (3.38)$$

$$\frac{3m}{N} = 0 \quad (3.39)$$

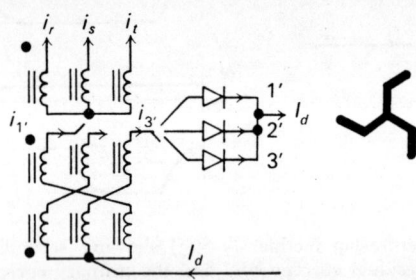

Fig. 28   Series-interconnected-secondaries with which
the residual m.m.f. of Fig. 27 has been cancelled

Parallel connection of two ($q = 3$) rectifiers must maintain 120° diode conduction to prevent the occurrence of $q = 6$ operation. This is achieved with an interphase reactor, the operation of which is described in Chapter 5.

A delta-connected primary or tertiary will suppress a triplen residual flux but will have no effect on a unidirectional residual flux. For the $q = 6$ rectifier, e_m is suppressed, keeping the secondary voltages sinusoidal, and the operation as described in the previous Section. As mode-1 operation will now predominate, the utilisation of secondary copper and diodes will be poorer than for a star-primary $q = 6$ rectifier which rapidly develops mode-2 operation with longer diode conduction angles and a form factor nearer unity.

46

3.4.2 Winding apparent-power (VA) ratings

The alternating supply current waveform to a centre-tap rectifier is determined by the m.m.f. balance condition from the device current waveforms (Figs. 26 and 27). The ratings of the transformer primary and secondary, assumed to have the same number of turns, are obtained from the r.m.s. value of their respective current waveforms. The secondary rating of a centre-tap-rectifier transformer is always greater than its primary rating owing to the presence of a component of direct current.

While leakage inductance slows up the current transfer at commutation, producing overlap, it is usual when calculating the r.m.s. of the current to assume that the current waveforms are rectangular (equivalent to zero leakage inductance). A high estimate of r.m.s. current is so obtained. The error is greatest for diode rectifiers where the overlap angles are not negligible: thyristors operating at any appreciable firing delay have much smaller overlap angles for which the rectangular assumption is valid.

The effect of firing-angle delay in a centre-tap rectifier without a freewheeling diode is to delay the entire current waveform with respect to the voltage by the angle of delay α. For a given output current I_d, the delay α only affects the shape of the a.c. waveform by progressively shortening the overlaps as α approaches 90°. The delay α does not therefore affect the transformer primary or secondary apparent-power ratings.

The output power of the rectifier, for a constant current I_d, falls off in proportion to the output voltage $V_d \cos \alpha$; and, assuming an efficiency of 100 % for the transformer and rectifier, the input power must also fall off proportionally. Whereas, by the above arguments, the primary apparent power is unaffected by the delay α, the primary current waveform is delayed with respect to the voltage waveform by the delay angle α, thus lowering the power factor of the rectifier. At $\alpha = 90°$ and with a constant current I_d, the output power is zero, and so is the primary-winding power factor. The concept of power factor for a rectifier is developed further in Chapter 5, where the effect of the nonsinusoidal current is also incorporated.

47

The relationships between the rectifier output power and the apparent-power ratings of the transformer primary and secondary windings are now derived for the more important centre-tap circuits. The leakage inductance of the transformer is neglected, so that the output voltage at rated power P_d is equal to the no-load voltage.

For the single-phase ($q = 2$) circuit, the secondary- and primary-current waveforms, obtained using m.m.f. balance, are illustrated in Fig. 29. The output power from previous relationships is

$$P_d = \frac{2E}{\pi} I_d \qquad (3.40)$$

where E is the peak secondary voltage.

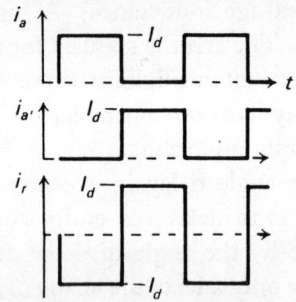

Fig. 29 Secondary and primary current for a single-phase ($q = 2$) rectifier, assuming zero leakage inductance

The r.m.s. value of the secondary current, flowing for half the total time, is $1/\sqrt{2}$ times the r.m.s. current during device conduction; i.e.

$$I_{ar} = I_d/\sqrt{2} \qquad (3.41)$$

whereupon

$$VA_{aa'} = \frac{2E}{\sqrt{2}} \frac{I_d}{\sqrt{2}} = I_d E \qquad (3.42)$$

The r.m.s. primary current I_r is equal to I_d, giving a primary apparent power

$$VA_r = \frac{E}{\sqrt{2}} I_d \qquad (3.43)$$

The relationships for the centre-tap circuit are thus

$$P_d:VA_{aa'}:VA_r = 1:\pi/2:\pi/2\sqrt{2}$$

$$= 1:1.57:1.11 \tag{3.44}$$

For the 3-phase centre-tap circuit ($q = 3$), working from Fig. 27, the output power is

$$P_d = \frac{3\sqrt{3}E}{2\pi} I_d \tag{3.45}$$

The secondary apparent-power rating is

$$VA_{abc} = 3\frac{E}{\sqrt{2}}\frac{I_d}{\sqrt{3}} \tag{3.46}$$

The primary apparent-power rating is

$$VA_{rst} = \frac{3E}{\sqrt{2}}\frac{\sqrt{2}I_d}{3} \tag{3.47}$$

giving
$$P_d:VA_{abc}:VA_{rst} = 1:\frac{\sqrt{2\pi}}{3}:\frac{2\pi}{3\sqrt{3}}$$

$$= 1:1.48:1.209 \tag{3.48}$$

For the diametrical 6-phase centre-tap circuit ($q = 6$), the results, based on Fig. 26, are

$$P_d:VA_{abc}:VA_{rst} = 1:1.812:1.047 \tag{3.49}$$

The increase in the secondary apparent-power rating resulting from reduced conduction angles (180°, 120° and 60°) is clearly shown in the results for the three circuits.

3.5 Reactance and regulation

The supply reactance, and the transformer leakage inductance, neglecting resistive voltage drops, are wholly responsible for the drop of rectifier output voltage as load is applied. As the transformer leakage reactance is frequently expressed as a percentage of the rated impedance, it is useful to have a direct relationship between the percentage reactance and the percentage regulation. When calculating this relationship, it must be remembered that the percentage react-ance is based on the r.m.s. values of sine waves, whereas the actual current waveforms of a rectifier transformer are nonsinusoidal.

Diode rectifiers only are considered in the following calculations of rectifier regulation. The actual voltage drop of controlled rectifiers is, however, the same; but their no-load voltage is reduced by the firing-angle delay. The relationships below all express the voltage drop as a percentage of the no-load voltage of a diode rectifier (= 100 %) or a controlled rectifier operating with zero firing delay.

The centre-tap rectifier-transformer has a secondary rating larger than its primary rating, owing to the component of direct current in the secondary winding. The rated direct output current I_d is used with the m.m.f. balance condition to calculate the rated r.m.s. primary current I_{rr}. If the transformer reactance is x per cent., then x per cent. of rated primary voltage is required to circulate rated primary current I_{rr} when the transformer secondary is short-circuited.

Applying this relationship to the single-phase centre-tap circuit, and assuming for simplicity a 1:1 turns ratio between primary and half secondary, we have

$$\frac{x}{100} \frac{E}{\sqrt{2}} = I_{rr}\{(\tfrac{1}{2})^2 \, 2\omega h\} = I_d \frac{\omega h}{2} \tag{3.50}$$

referring the total secondary leakage inductance to the primary. Com-bining this with the output-voltage relationship

$$V_d = \frac{2E}{\pi} - \frac{I_d \omega h}{\pi} \tag{3.51}$$

we have

$$\frac{100 V_d}{2E/\pi} = 100 - 100 \frac{I_d \omega h}{2E} = 100 - \frac{x}{\sqrt{2}} \tag{3.52}$$

from which it is clear that a transformer with x per cent. reactance gives a rectifier with $x/\sqrt{2}$ per cent. voltage drop at full load.

The 3 phase centre tap rectifier has a rated r.m.s. primary current of $I_d\sqrt{2/3}$. Taking for simplicity a 1:1 star–star transformer, and considering one phase only,

$$\frac{x}{100} \frac{E}{\sqrt{2}} = \omega h I_d \sqrt{2/3} \tag{3.53}$$

Combining this with the voltage relationship

$$V_d = \frac{3\sqrt{3}E}{2\pi} - \frac{3I_d\omega h}{2\pi} \tag{3.54}$$

we have
$$\frac{100V_d}{3\sqrt{3}E/2\pi} = 100 - 100\frac{I_d\omega h}{\sqrt{3}E} = 100 - \frac{x\sqrt{3}}{2} \tag{3.55}$$

The 6-phase (diametrical) star has a rated primary current of $I_d\sqrt{2/3}$. Again taking a $1:1$ star–star transformer, and considering one phase only,

$$\frac{x}{100}\frac{E}{\sqrt{2}} = (\tfrac{1}{2})^2\, 2\omega h(I_d\sqrt{2/3}) \tag{3.56}$$

Combining this with the voltage relationship

$$V_d = \frac{3E}{\pi} - \frac{3I_d\omega h}{\pi} \tag{3.57}$$

we have
$$100\frac{V_d}{3E/\pi} = 100 - 100\frac{I_d\omega h}{E} = 100 - 3x \tag{3.58}$$

It is clear that the regulation becomes worse as q increases (for mode-1 operation, on which the above calculations are based), bearing out the regulation curves of Fig. 23.

4 BASIC RECTIFIER CIRCUITS: BRIDGE RECTIFIERS

4.1 Introduction

The same assumptions and introductory remarks apply equally to bridge rectifiers as to centre-tap rectifiers.

Bridge rectifiers are essentially fullwave: they have rectifying devices between each terminal of the load and each terminal of the a.c. supply, as shown generally in Fig. 30. For q a.c. terminals, there are thus $2q$ devices. For a given phase voltage $e = E\sin\omega t$, the device voltage rating and rectifier output-current rating are the same as for

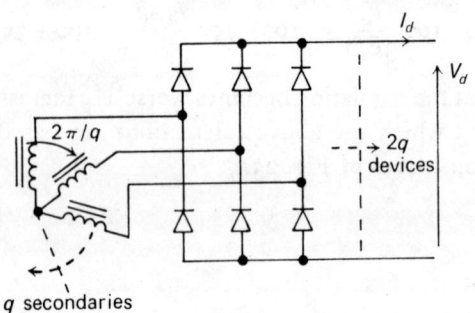

Fig. 30 Generalised bridge rectifier having q secondaries and $2q$ devices

the centre-tap arrangement, but the output voltage V_d is doubled, at the expense of having double the number of devices. In fact, the circuit can be considered as two centre-tap rectifiers of opposite polarity in series, but with no external connection to the (common) centre tap.

For odd values of q (except $q = 1$, which does not constitute a bridge, having only two diodes), the overlaps on the positive and negative sides of a diode bridge occur alternately, giving an output ripple frequency $f_r = 2qf_s$, where f_s is the supply frequency. But, for even values of q, the positive- and negative-side overlaps occur together, so that the ripple frequency $f_r = qf_s$. The most common values of q are 2 and 3, for single- and 3-phase supplies. Higher values of q are rarely used because of poorer device and transformer utilisation.

In this treatment of bridge rectifiers, the transformer secondaries are always shown in the centre-tap (star) arrangement, the potential of the centre tap being the datum of the voltage waveforms. No electrical connection is made to the centre tap; it is merely a graphical convenience. In practice, delta secondaries are equally acceptable, or, for that matter, the direct connection of a bridge rectifier to a supply.

Unlike the centre-tap rectifiers, which have the transformer centre tap as an output terminal, bridge rectifiers have both output terminals fed from device terminals connected in common. To avoid confusion, the output terminals of bridge rectifiers will not be described as positive or negative, since each can assume either polarity depending on whether the bridge is rectifying or inverting. Instead, the descriptions 'cathode output' (c.o.) and 'anode output' (a.o.) will be used to distinguish between the output terminals or their potentials, according to whether the devices have their cathodes or anodes connected in common to the output terminal. For a diode bridge, the cathode-input terminal follows the phase waveform with the most positive potential, and the anode-output terminal follows that with the most negative. The output voltage V_d is the potential difference.

Bridge rectifiers have the marked advantage over centre-tap rectifiers in that the phases supplying the bridge do not carry a component of direct current. For a given device conduction angle, a cycle of phase current is composed of two such periods of current in opposite directions, carried by the two connected devices alternately. The absence of the d.c. component reduces the rating of the transformer secondary to equality with the primary, and, where the voltages are suitable, allows the use of a rectifier without a transformer.

Controlled bridge rectifiers can be either fully or half controlled, depending on whether all or half of the devices are thyristors. A half-controlled bridge (Freris, 1966) has its thyristors connected to one

53

output terminal and its diodes connected to the other. The single-phase case has an alternative (asymmetrical) arrangement in which the two thyristors are common to an a.c. terminal. Fully controlled bridges have the ability to invert, which is not possessed by half-controlled bridges.

The analysis of bridge rectifiers follows the same pattern as for centre-tap rectifiers; the voltage–time area for a repetition period of output voltage is divided by the period to obtain the mean output voltage. The voltage–time areas lost in overlap and phase delay are obtained and subtracted as before.

4.2 Input–output parameters: single-phase
4.2.1 Fully controlled bridge

In the diode bridge, diagonally opposite diodes conduct together to establish a circuit between the a.c. supply and d.c. output. During overlap, the bridge symmetry usually present demands that all four diodes conduct, giving zero voltage output and a momentary short circuit on the transformer secondary. A conduction pattern of 2–4–2–4 etc. therefore applies to the diode-bridge rectifier.

Where all the devices are thyristors, the bridge is fully controlled, whereas, if only 50% are thyristors, the bridge is half controlled. Both forms are important, and will be examined in turn. It is usual to fire both incoming thyristors of a fully controlled bridge together, implying that the delay angles for the upper and lower half bridges are equal. This is not an essential restriction, and bridge operation for unequal delay angles is examined towards the end of the Section. Finally, the use of a single thyristor in series with the output of a diode bridge is considered.

The operation of a fully controlled single-phase bridge, which includes the operation of a diode bridge by setting $\alpha = 0$, is shown in Fig. 31, in which the ringed numbers define the conducting devices. During overlap, all four thyristors are conducting, giving zero output, as shown by the c.o. terminal potential (boundary shaded below the line) and the a.o. terminal potential (boundary shaded above the line) being equal at the potential of the mean of the overlapping phases, i.e. at the centre-tap potential for this circuit. The output voltage V_d is

$$V_d = \frac{\omega}{\pi}\left(\frac{4E}{\omega} - \frac{4E}{\omega}\,2\sin^2\frac{\alpha}{2} - 2I_d 2h\right) \qquad (4.1)$$

The initial voltage–time area $4E/\omega$ is clearly double that of the centre-tap circuit; the delay area is four times that shown in Fig. 7 since negative area has been gained and positive area lost. The overlap

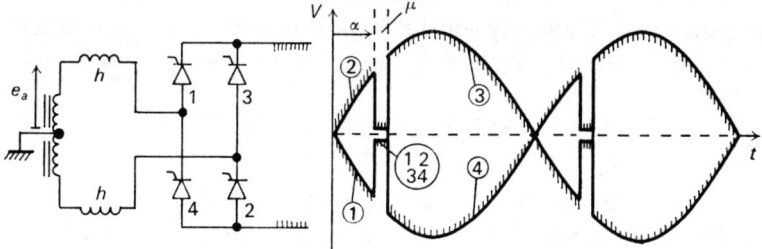

Fig. 31   Single-phase bridge rectifier

Circuit diagram, and potentials of anode-output (a.o.) and cathode-output (c.o.) terminals boundary shaded above and below the line. Circled numbers indicate conducting devices

area results from the reversal of I_d in a total inductance of $2h$. V_d can alternatively be expressed as

$$V_d = \frac{4E}{\pi}\cos\alpha - \frac{4I_d\omega h}{\pi} \qquad (4.2)$$

The overlap angle α is most easily obtained by considering the current reversal in the a.c. phases during overlap when the bridge acts as a short circuit:

$$2I_d 2h = \frac{2E}{\omega}\left(2\sin^2\frac{\alpha+\mu}{2} - 2\sin^2\frac{\alpha}{2}\right) \qquad (4.3)$$

Asymmetry in the bridge is now briefly considered to show its effect on overlap and demonstrate that the subsequent analysis for differing firing-angle delays for the upper and lower halves of the bridge can be applied to it. Fig. 32 shows a bridge in which the thyristors 1 and 4 have a forward drop of 3 V, while thyristors 2 and 3 have a forward

55

drop of 1 V. Thyristors 1 and 2 have been conducting, and, when 3 and 4 are fired, $e_a = -e$ and $e_b = e$ with respect to the centre tap. As no current flows from the centre tap, di/dt in both half secondaries is equal, and hence the voltage drops across the inductances h are also equal.

On firing 3 and 4, 3 will immediately establish a freewheeling path of only 2 V drop through 2 and 3; hence, although 4 is fired, it does not experience a forward voltage, by the following argument:

As 1 and 3 are both conducting, V_{14} must be 2 V positive with respect to V_{32}. To satisfy equal voltage drops across the two inductances, the potentials V_{14} and V_{32} must be $+1$ V and -1 V, respec-

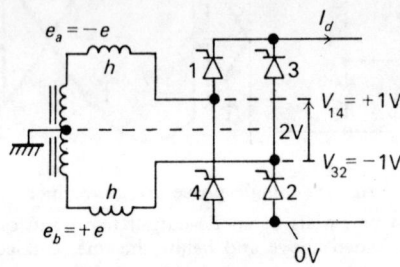

Fig. 32 Showing how differences in device forward-voltage drop results in sequential rather than simultaneous commutation of upper and lower devices Forward drop of 1 and 4 = 3 V, of 2 and 3 = 1 V. Thyristors 1 and 2 have been conducting previously. Instantaneous potentials are shown at the instant when 3 and 4 are fired; 3 conducts but 4 remains reverse-biased until the 1–3 commutation is complete

tively. But as 2 is also conducting, the potential of the a.o. terminal must be 0 V, and hence 4 is reverse-biased with 1 V, whereas it requires 3 V of forward bias for it to begin conduction. Thus the rate of change of phase current $(2e+2)/2h$ is satisfied wholly by a rise of current in 3 and a collapse of current in 1; i.e. the upper devices 1 and 3 commutate while 2 carries I_d steadily. Only when the upper commutation is complete will the lower commutation begin, resulting in a conduction pattern of 2–3–3–2–3–3 etc. The upper commutation persists while the phase current collapses from $+I_d$ to zero, while the lower commutation proceeds as the phase current builds up from zero to $-I_d$. The total voltage–time area lost during both overlaps is

56

the same as before, and eqns. 4.1 and 4.2 apply. The total overlap μ is still given by eqn. 4.3, but is made up to two overlaps, each separately calculable from the corresponding change of current $I_d 2h$. Thus the behaviour of a bridge with unequal forward voltage drops can be considered as a bridge operating with unequal firing delays on its two halves.

4.2.2 Half-controlled bridge

The half-controlled bridge is an important single-phase circuit, since voltage control over the entire range is possible for only 50% thyristors. Inversion is, however, not possible.

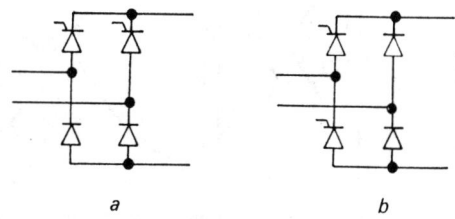

Fig. 33   Symmetrical (a) and asymmetrical (b) single-phase
bridge arrangements
The latter possesses an inbuilt freewheeling path through the two diodes

While the behaviour of a half-controlled bridge is a special case of the bridge with unequal firing delays on its two halves, it is treated separately here because of its importance. Two forms of half-controlled bridge exist for single phase; they are shown in Figs. 33a and b. The first, with the thyristors common to a d.c. terminal (either common anodes, or common cathodes as shown), is more frequently used than the second, which has its thyristors common to an a.c. terminal. However, the latter circuit has an inbuilt freewheeling path not possessed by the former.

The behaviour of the circuit of Fig. 34 is considered first. The diodes commutate at the beginning of each halfcycle, giving a period of zero output voltage during their overlap. A thyristor fired during this overlap will not alter the bridge condition, as the voltage across the a.c. terminals of the bridge is already zero as a result of the diode commutation. There may even be insufficient voltage for the incoming thyristor to fire correctly, in which case the thyristor firing pulse must have a duration longer than the diode overlap, so that, when the phase voltages are released from the diode overlap, a firing pulse is still present on the incoming thyristor to initiate conduction. The waveforms for a delay α are shown in Fig. 34, the ringed numbers indicating the conducting devices as usual. The overlap

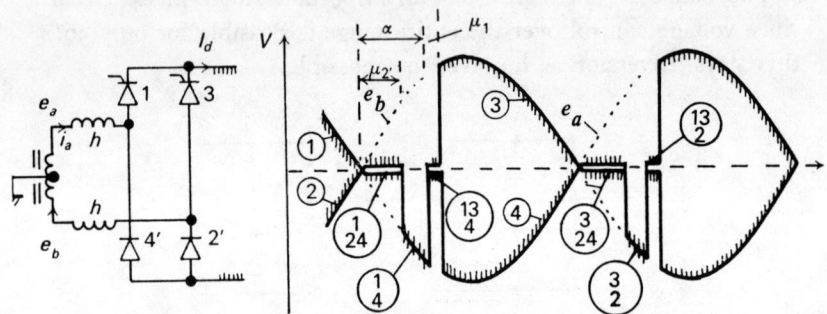

Fig. 34   The half-controlled single-phase bridge showing the output potentials of the c.o. and a.o. terminals

Ringed numbers indicate conducting devices. Although the overlap angles $\mu_{2'}$ and μ_1 are unequal, their voltage–time areas are equal

angles for diodes and thyristors are shown as $\mu_{2'}$ and μ_1. The output voltage $V_{d\alpha}$ is

$$V_{d\alpha} = \frac{\omega}{\pi}\left(\frac{4E}{\omega} - \frac{2E}{\omega}2\sin^2\frac{\alpha}{2} - 2I_d\ h\right) \tag{4.4}$$

$$= \frac{4E}{\pi}\left(\tfrac{1}{2} + \tfrac{1}{2}\cos\alpha\right) - \frac{2I_d\omega h}{\pi} \tag{4.5}$$

Eqns. 4.4 and 4.5 only apply for $\alpha > \mu_{2'}$, as values of α less than this do not raise the output voltage any more, owing to the unavoidable loss of voltage–time area during diode overlap, which has not now been separately included in eqns. 4.4 and 4.5. The device and phase-current waveforms are easily derived from the conduction of the devices, as shown in Fig. 35.

Fig. 34 shows clearly the freewheeling action that takes place after the diode overlap but before the firing of the incoming thyristor. The freewheeling current flows through a thyristor (rated accordingly), which reveals one advantage of the other half-controlled bridge arrangement (Fig. 33 b), in which the freewheeling always takes place through the two diodes, thus relieving the thyristors of this current.

The alternative arrangement for the half-controlled bridge (Fig. 36) is analysed in a similar manner. The output-voltage and the line-current waveforms are the same for both circuits, so that eqns. 4.4

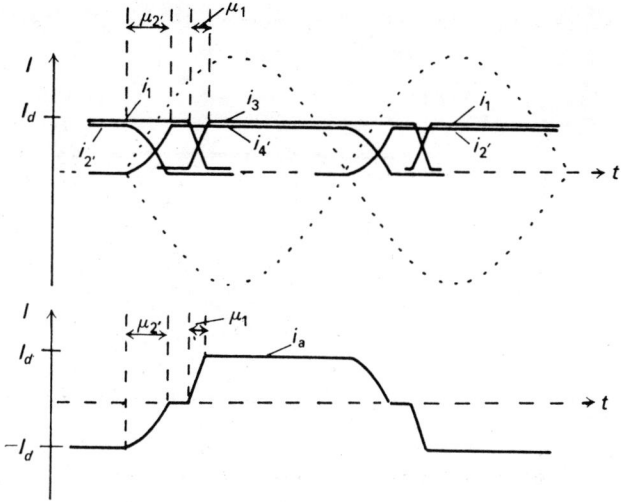

Fig. 35   Device current and alternating-supply current for Fig. 34

and 4.5 apply. As the delay is increased, the thyristor conduction angles decrease, while the diode conduction angles increase. For a smooth direct current I_d, the reduction of thyristor conduction angle is of little consequence, since the thyristors must be rated for I_d at 180° conduction angle to meet the maximum-output condition. A discontinuous output current, which often develops at lower output voltages for low load inductance, can mean that a peak output current considerably greater than I_d (mean) is required to give a mean output current of I_d. This throws an additional load on the thyristors of Fig. 34; this load is less severe for the thyristors of Fig. 36, because of their decreased conduction angle.

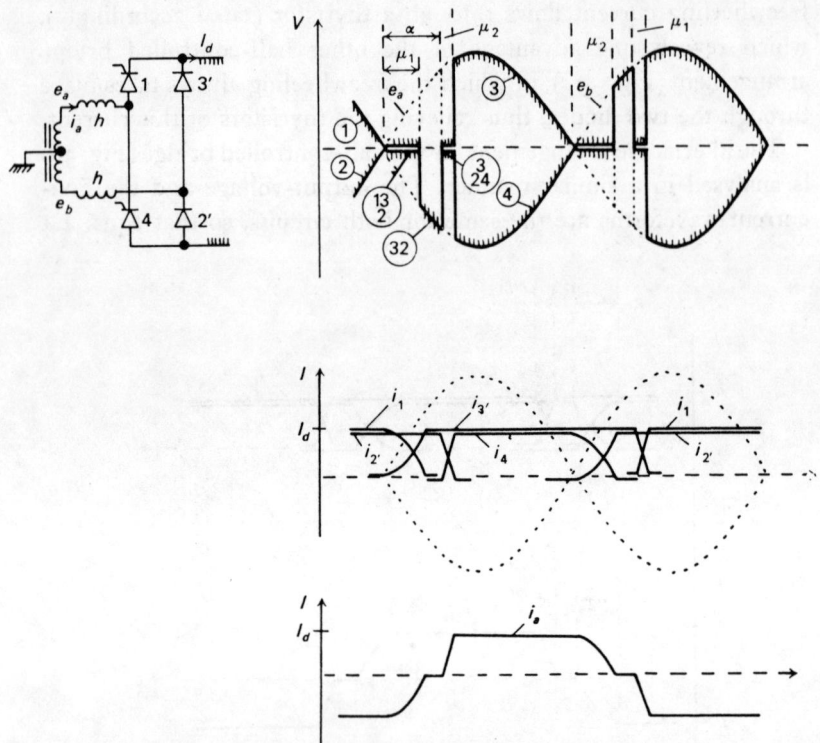

Fig. 36 Voltage and current for asymmetrical single-phase
half-controlled bridge

Note the shorter conduction periods for the thyristors
when compared with Figs. 34 and 35

4.2.3 Unequal delay angles α_1 and α_2

The most general case of unequal delays α_1 and α_2 for the upper
and lower halves of the single-phase bridge is now considered. The
waveforms are illustrated in Fig. 37, which also indicates the conducting thyristors by ringed numbers as before. Anode and cathode
output-terminal potentials are boundary shaded above and below the
lines to make clear the output voltage–time area over one repetition
period; this area is shaded vertically where positive and horizontally

where negative. The output voltage is

$$V_d = \frac{\omega}{\pi}\left(\frac{4E}{\omega} - \frac{2E}{\omega} 2\sin^2 \frac{\alpha_1}{2} - \frac{2E}{\omega} 2\sin^2 \frac{\alpha_2}{2} - I_d 2h\right) \qquad (4.6)$$

$$= \frac{2E}{\pi}\cos\alpha_1 + \frac{2E}{\pi}\cos\alpha_2 - \frac{2I_d\omega h}{\pi} \qquad (4.7)$$

Eqns. 4.6 and 4.7 reduce to eqns. 4.5 and 4.6 for $\alpha_2 = 0$, provided also that $\alpha_1 \nless \mu_2$, a restriction which applies as before, since the overlap μ_2 collapses the phase voltages to zero and thus prevents a delay $\alpha_1 < \mu_2$ from raising the output voltage further.

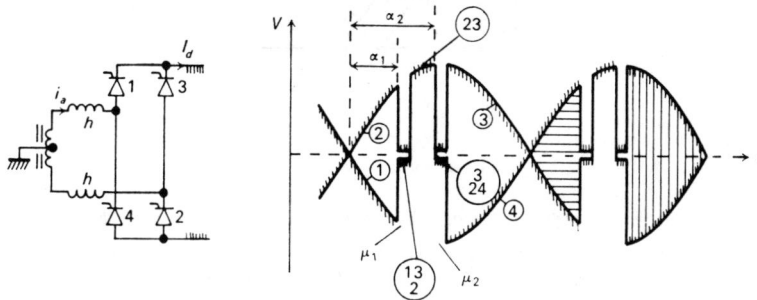

Fig. 37   Voltage waveforms and conducting devices (ringed)
for a fully controlled bridge with unequal delays α_1 and α_2
Positive output voltage–time area is shaded vertically, and negative
area is shaded horizontally

Overlap interference occurs in its most general form when the delay angle α_1 is progressively increased, so that $\alpha_1 + \mu_1 > \alpha_2$, continuing until $\alpha_1 > \alpha_2 + \mu_2$ when separate overlaps again occur.

At first, increasing α_1 so that $\alpha_1 + \mu_1 > \alpha_2$ merely delays the start of μ_2, so that the rectifier behaviour can be represented by both α_1 and α_2 being increased together. The output voltage consequently falls off at double the rate during this first stage.

When $\alpha_1 = \alpha_2$, the two overlaps μ_1 and μ_2 can be considered to take place in either order; but, for $\alpha_1 > \alpha_2$, μ_2 will occur first, so that the incoming thyristor 1 or 3 will be fired during the overlap μ_2 and consequently will not fire or alter the waveforms. Increasing α_1 from

α_2 to $\alpha_2 + \mu_2$ has no effect on the output voltage, as μ_1 cannot start until the end of μ_2, whatever the value of α_1 in this range.

The way in which V_d varies with α_1 for two values of α_2 is shown in Fig. 38.

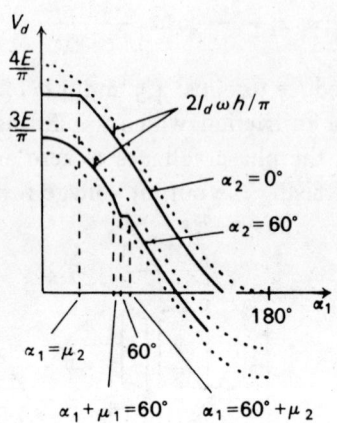

Fig. 38   Control characteristics for α_1 when $\alpha_2 = 0$ and $\alpha_2 = 60°$
Overlap interference is shown by the regions where the control
characteristic departs from a smooth curve

4.2.4 Single-thyristor bridge

An important single-phase circuit requiring only one thyristor is shown in Fig. 39. It is possible only for single phase, as its operation depends on the short period between the two halfcycles of the bridge output voltage for thyristor turnoff. A thyristor with a short turnoff time is usually required (Durnya, 1968) for reliable operation, which is also assisted by the capacitor (shown dotted) retaining a small positive voltage on the thyristor cathode, at the end of each halfcycle.

At maximum output, the thyristor is fired as early as possible in each halfcycle, and it starts to conduct at the end of the diode overlap when the bridge output voltage rises from zero. For reduced output, the thyristor is fired later; overlap of the diodes will only proceed as far as zero phase current, while the freewheeling diode current builds up to maintain I_d constant. When the thyristor is fired, the current I_d

must be re-established in the leakage inductance $2h$, so that an overlap now occurs between the freewheeling diode and the thyristor, the two appropriate bridge diodes conducting to complete the circuit. A voltage–time area $2I_d h$ is therefore lost after firing the thyristor. One can consider the behaviour conveniently using the equivalent circuit of Fig. 40, in which the equivalent source has the output voltage waveform of the bridge rectifier at no load.

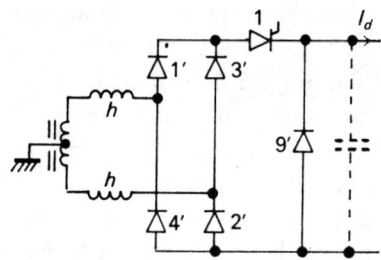

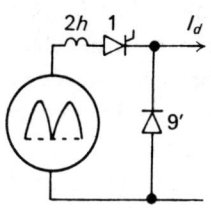

Fig. 39 Single thyristor-controlled recti-fier with freewheeling diode, suitable for an inductive load

Fig. 40 Equivalent cir-cuit for Fig. 39 in which the source-voltage wave-form is the no-load output-voltage waveform of the uncontrolled rectifier

4.3 Input–output parameters: 3-phase

The 3-phase bridge ($q = 3$, Graetz bridge) is the most frequently used high-power rectifier circuit in its uncontrolled, half-controlled or fully controlled form. The output voltage of an uncontrolled bridge has a mean ripple voltage at light load of only $3 \cdot 6 \%$ of its mean output voltage, with a ripple frequency $f_r = 6f_s$, and the output is thus a good approximation to a smooth direct voltage. Like all bridge rectifiers, it only requires a rectifier transformer when the required output voltage cannot be obtained by the direct rectification of the main supply, or where the d.c. output must be isolated from the supply, or where the high power rating demands that the rectifier is connected to a higher-voltage distribution feeder. The 3-phase bridge is widely used in the power range 10–500 kW for supplying he armature of a variable-speed d.c. motor; the transformer is often

63

omitted, and the armature is designed for the voltage resulting from the direct rectification of the a.c. supply.

This analysis retains the transformer on the understanding that, where it is not used, the supply reactance, with or without added reactors as appropriate, takes the place of the transformer leakage reactance.

The six diodes in a 3-phase bridge start conducting in a fixed sequence at equal time intervals. They are numbered 1'–6' in this sequence. In a half-controlled rectifier, however, a firing delay of more than 60° for the thyristors 1, 3 and 5 will result in the firing of 3 after the start of conduction of diode 4' on the other side of the bridge. The device numbering is not altered to match these conditions, but remains as applicable to the zero-firing-delay behaviour. To meet the essential condition of balanced 3-phase operation, for which the three line-current waveforms are identical except for 120° phase displacement, it is only necessary that the intervals between firing the devices 1, 3 and 5 are 120° and the intervals between firing the devices 2, 4 and 6 are also 120°.

4.3.1 Fully controlled bridge

The fully controlled bridge with equal delays on all thyristors is considered first [Adamson and Hingorani, 1960; Cory (Ed.), 1965]: the results apply also to the diode bridge on substituting $\alpha = 0$ (Ludbrook and Murray, 1965). The circuit and waveforms are given in Fig. 41. The thyristor numbering indicates the conduction sequence, and the interval is 60°. The conduction pattern for simple overlap conditions is 2–3–2–3, the overlap between 1 and 3 occurring while the current I_d returns through 2. The next overlap occurs between 2 and 4 while the current I_d flows through 3. During overlap, the potential of the bridge output terminal connected to the overlapping thyristors follows the mean of the overlapping phase voltages, giving the line and output waveforms shown in Fig. 41. From these waveforms, the output voltage V_d is written down as

$$V_d = \frac{3\omega}{\pi}\left(\frac{\sqrt{3}E}{\omega} - \frac{\sqrt{3}E}{\omega}2\sin^2\frac{\alpha}{2} - I_d h\right) \qquad (4.8)$$

giving
$$V_d = \frac{3\sqrt{3}E}{\pi}\cos\alpha - \frac{3I_d\omega h}{\pi} \tag{4.9}$$

The overlap μ is given by

$$I_d h = \frac{\sqrt{3}E}{2\omega}\left(2\sin^2\frac{\alpha+\mu}{2} - 2\sin^2\frac{\alpha}{2}\right) \tag{4.10}$$

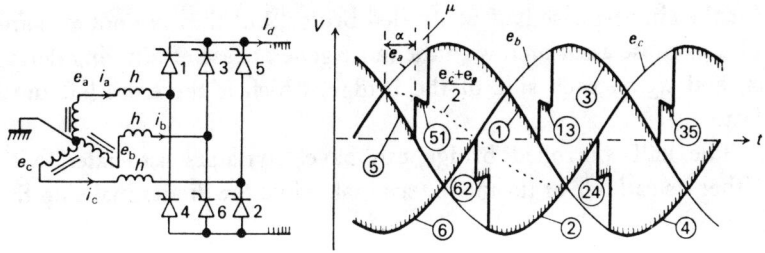

Fig. 41 Circuit and output-terminal potential for fully controlled
3-phase bridge in which all thyristors have same delay angle α

Ringed numbers indicate which thyristors are conducting

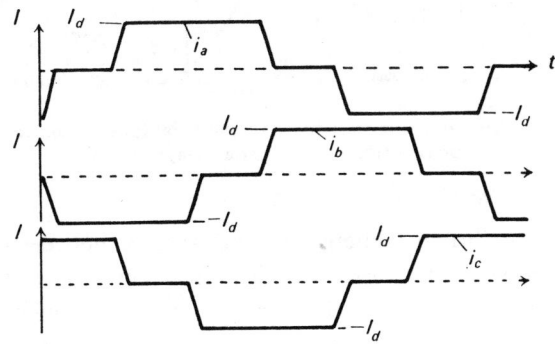

Fig. 42 Alternating-current waveforms for the 3-phase bridge of Fig. 41

It is clear from Fig. 41 that the devices have a conduction angle of
$120° + \mu$, and, as each a.c. phase supplies two devices, the phase
currents will be as shown in Fig. 42.

4.3.2 Half-controlled bridge

The half-controlled bridge is considered next, as it is a commonly used arrangement in the power range 5–100 kW, for full-range voltage control without inversion (Freris, 1966; Duff and Ludbrook, 1968). Some problems arise from overlap interference as they did for the single-phase half-controlled bridge, but they are not as complex as those associated with the more general case of differing delays α_1 and α_2 on each side of the bridge, which is therefore left until last.

The half-controlled bridge can have thyristors associated with either its cathode or its anode terminal, while the diodes make up the

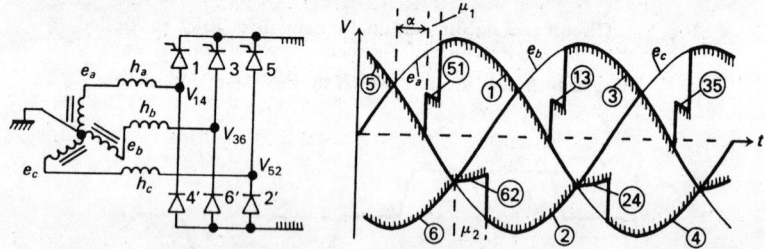

Fig. 43   Circuit and output-terminal potential for half-controlled
3-phase bridge for thyristor delay angle α

other half of the bridge. Where electrical isolation between gates and control circuits is not necessary, it is convenient to have thyristors connected to the cathode output terminal, as the common cathodes allow the gates to be fed directly from the control circuitry, avoiding pulse transformers. This is the arrangement illustrated in Fig. 43. The diodes operate as in a centre-tap rectifier, the anode output terminal following the most negative of the three phase potentials except during overlap, when it follows the mean of the overlapping phase potentials. The thyristors, fired after a delay α, give rise to the anode output-terminal potential shown. The voltage–time areas lost in overlaps between diodes or thyristors, while differing in shape, are equal in area ($= I_d h$). The output voltage is, for a repetition period $2\pi/3\omega$,

$$V_d = \frac{3\omega}{2\pi} \left(\frac{2\sqrt{3}E}{\omega} - \frac{\sqrt{3}E}{\omega} 2\sin^2\frac{\alpha}{2} - 2I_d h \right) \qquad (4.11)$$

i.e. $$V_d = \frac{3\sqrt{3}E}{\pi} (\tfrac{1}{2} + \tfrac{1}{2}\cos\alpha) - \frac{3I_d\omega h}{\pi} \qquad (4.12)$$

the latter formula exhibiting the usual behaviour of half-controlled rectifiers in the control term ($\tfrac{1}{2} + \tfrac{1}{2}\cos\alpha$).

Whereas for the single-phase bridge ($q = 2$), and for even values of q, overlap interference occurs at $\alpha = 0$ in half-controlled rectifiers, for the 3-phase bridge ($q = 3$) and ($q = 2n+1$), the overlaps on the upper and lower halves occur alternately, so that no interference occurs for $\alpha = 0$. Instead, interference occurs at $\alpha = 60°$ and $180°$ for the 3-phase bridge. Unusual waveforms develop where there is overlap interference, which are reminiscent of multiple overlaps for centre-tap circuits. In particular, the buildup of diode current, followed almost immediately by its collapse to zero after firing a thyristor, give rise to diode current waveforms somewhat similar to those of Fig. 20; the voltage–time areas which bring about such current changes are also similar.

Overlap interference begins when $\alpha + \mu_1 > 60°$, and ends when $\alpha > 60° + \mu_2$. As for the single-phase half-controlled bridge, the first effect of overlap interference is for the thyristor overlap μ_1 to delay the start of the diode overlap μ_2. Increasing α thus has an effect of delaying the diode commutation also, so that the output voltage falls off slightly more rapidly than changes in α alone would produce. The waveforms illustrating this region of interference are given in Fig. 44.

The second region of interference occurs for a delay α slightly greater than $60°$. The resulting waveforms are shown in Fig. 45. The diode overlap begins normally at the phase crossover, when e_c becomes negative with respect to e_b. The current $i_{2'}$ builds up as $i_{6'}$ reduces until, at α, 1 is fired, which initiates a period during which four devices conduct together. This results in both output terminals taking the potential of the centre tap, until the vertically shaded area,

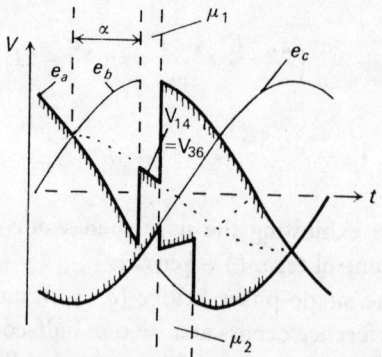

Fig. 44   Overlap interference when $\alpha + \mu_1 > 60°$

The thyristor commutation μ_1, involving e_a and e_b, deflects V_{14} which delays the start of the diode commutation μ_2 until the completion of μ_1

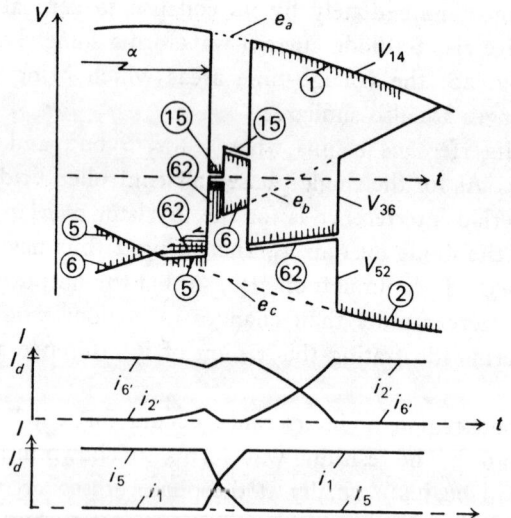

Fig. 45   Overlap interference when $\alpha + \mu_1 > 60°$

Horizontally and vertically shaded areas are equal. The period when devices 1, 5, 6′ and 2′ conduct together results in a momentary 3-phase short circuit, with both output potentials falling to zero (neutral) potential

68

which is increasing $i_{6'}$, is equal to the horizontally shaded area, which previously reduced $i_{6'}$ during the initial diode overlap.

It is easier to focus attention on diode 6′ and the current changes brought about in h_b, as the diode 6′ is the only conducting device connected to h_b. In contrast, h_c is connected to diode 2′ and thyristor 5, both of which participate in the overlaps, and hence changes of current in h_c can be met by changes in $i_{2'}$ or $i_{5'}$, or both.

The current $i_{2'}$ is readily obtained from $i_{6'}$, since $i_{2'} + i_{6'}'' = I_d$. When $i_{2'} = 0$, the diode 2′ ceases conduction, leaving only diode 6′ conducting in the lower-half bridge and thyristors 5 and 1 overlapping in the upper-half bridge. It is the 5–1 overlap which has deflected the phase V_{14} to a potential more positive than V_{36}, thereby stopping the diode overlap.

When the voltage–time area required to establish I_d in h_a has elapsed, the thyristor overlap terminates, releasing $V_{52'}$, hence immediately restarting the diode overlap 6′2′, which continues to completion normally.

As the delay α is increased, the current in the outgoing diode 6′ decreases further, and therefore a larger voltage–time area is required to build it up to I_d again. A critical condition is reached (Fig. 46) in which the horizontally shaded area indicates the area which reduces $i_{6'}$, and the vertically shaded area builds up $i_{6'}$. The critical condition for which i_6 just reaches I_d (or $i_{2'}$ just reaches zero) at the completion of the overlap μ_1 is defined by the equality of the two shaded areas when the second area has an angular duration μ_1.

$$\frac{\sqrt{3}E}{2\omega} 2\sin^2\frac{\alpha-60}{2} = \frac{E}{\omega}\left(2\sin^2\frac{90-\alpha}{2} - 2\sin^2\frac{90-\alpha-\mu_1}{2}\right)$$

$$\tag{4.13}$$

where μ_1 is given by

$$\frac{E}{\omega}\left(2\sin^2\frac{150-\alpha}{2} - 2\sin^2\frac{150-\alpha-\mu_1}{2}\right) = I_d h \tag{4.14}$$

The two simultaneous eqns. 4.13 and 4.14 give α and μ_1 for the critical condition in terms of E, ω, I_d and h.

Values of α larger than the critical value result in another region of overlap interference, for which the current waveforms are given in Fig. 47, the voltage waveforms remaining similar to those of Fig. 46. The diode commutation is slightly less than half completed when the

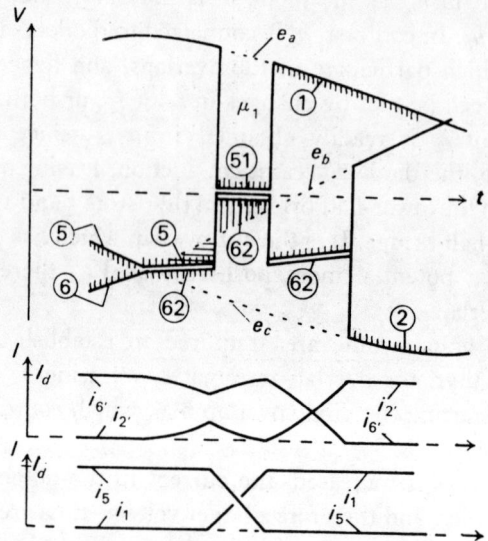

Fig. 46 Overlap interference when $\alpha + \mu_1 > 60°$
The critical condition applies when $i_{6'}$ just returns to I_d
at the completion of the thyristor overlap

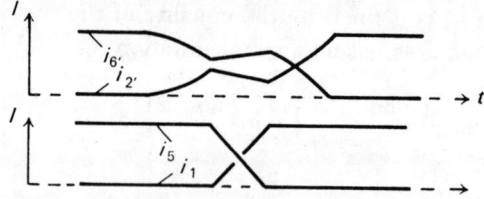

Fig. 47 Device current for a delay angle α greater
than the critical value illustrated in Fig. 46

thyristor is fired. Fig. 48 illustrates the waveforms for a diode commutation which is very nearly completed when the thyristor is fired, representing the final stage of overlap interference.

A further increase in α results in the two overlaps becoming

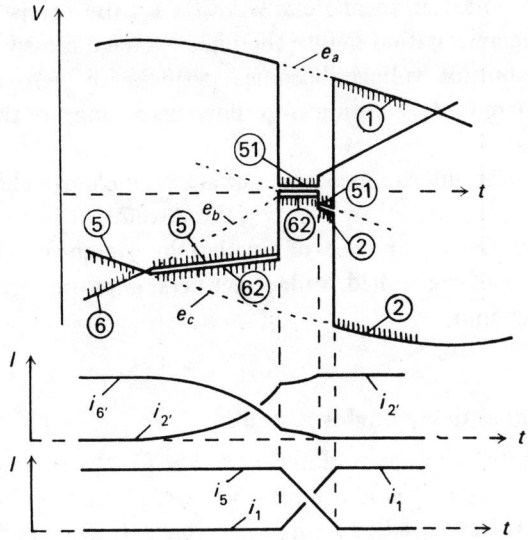

Fig. 48   The last stage of overlap interference
Diode overlap 6′2′ is almost complete when 1
is fired to initiate the 5–1 overlap

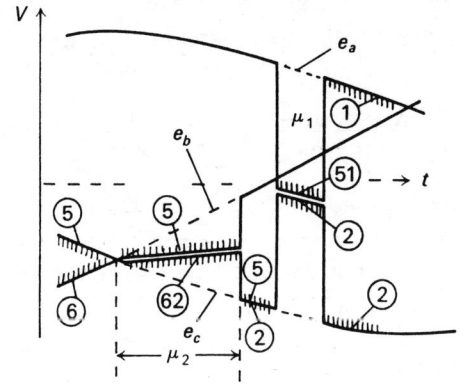

Fig. 49   Separate overlap, occurring when $\alpha > 60 + \mu_2$

separated, giving rise to waveforms shown in Fig. 49. It should be
noted that a period of 2-device freewheeling is now occurring through
2′ and 5, and further increase of α extends the freewheeling period

71

accordingly, until, at an angle $\alpha = 180° - \mu_2$, the thyristors cannot complete a commutation before the phase-voltage crossover, so that the bridge output voltage becomes permanently zero (while the constant current I_d continues to flow according to the original assumptions).

While overlap interference does not alter the control characteristic (V_d against α) by more than 1 or 2%, the waveforms are considerably modified by the occurrence of overlap interference, which, particularly for half-controlled bridge rectifiers, is an important aspect of their behaviour.

4.3.3 Unequal delay angles α_1 and α_2

The 3-phase bridge with differing delays for the thyristors of the upper and lower halves is the most generalised mode of operation, although the least used. The half-controlled bridge, and the fully controlled bridge with equal delays, are both widely used special cases.

Where overlap interference does not occur, no special problems exist, and the output voltage can be given either from a generalised form of Fig. 43 or from the operation of two series-connected centre-tap circuits:

$$V_d = \frac{3\omega}{2\pi}\left(\frac{2\sqrt{3}E}{\omega} - \frac{\sqrt{3}E}{\omega}2\sin^2\frac{\alpha_1}{2} - \frac{\sqrt{3}E}{\omega}2\sin^2\frac{\alpha_2}{2} - 2I_dh\right) \quad (4.15)$$

$$= \frac{3\sqrt{3}E}{\pi}\left(\tfrac{1}{2}\cos\alpha_1 + \tfrac{1}{2}\cos\alpha_2\right) - \frac{3I_d\omega h}{\pi} \quad (4.16)$$

Where overlap interference does occur, the behaviour cannot readily be generalised into a pattern which applies for all combinations of α_1 and α_2. Two examples which illustrate and contrast behaviour will be considered briefly.

When $\alpha_2 + \mu_2 < 90°$, the overlap potential, which is also the potential of the incoming phase for the upper-half bridge during μ_2, is negative with respect to the phase supplying the cathode output terminal (Fig. 50). Consider, first, $\alpha_1 < \alpha_2 - 60°$: the overlap μ_1 will begin at α_1, and, at α_2, four devices will conduct together, giving a behaviour similar to the half-controlled bridge (Fig. 45). Secondly, consider the case when α_1 occurs within μ_2: the incoming phase is no longer positive with respect to the conducting phase, so that the incoming thyristor will not fire at α_1. With a prolonged gate pulse, it will fire after μ_2, but variations of α_1 within μ_2 will clearly have no effect on the output voltage. The voltage control characteristic will thus have a horizontal portion when α_1 lies within μ_2, similar to that occurring with the single-phase bridge in Fig. 38.

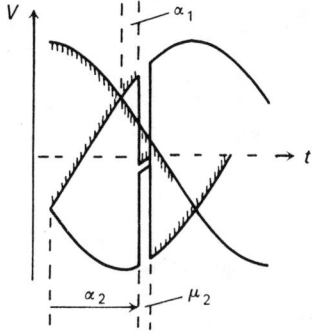 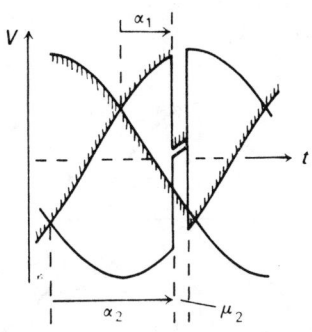

Fig. 50   Voltage waveforms for unequal delays α_1 and α_2
When $\alpha_2 + \mu_2 < 90°$, the incoming upper thyristor is reverse-biased if α_1 falls within μ_2

Fig. 51   Voltage waveforms for unequal delays α_1, α_2
When $\alpha_2 + \mu_2 > 90$, the incoming upper thyristor is always forward-biased

For $\alpha_2 > 90°$, the potential of the overlapping phases is positive with respect to the other phase potential, so that the incoming thyristor fired at $\alpha_1 > 30°$ will always be forward-biased, and will begin conduction at α_1 as shown in Fig. 51. Any value of α_1 which gives rise to overlap interference initiates a period of four thyristors conducting together, resulting in behaviour similar to Figs. 45–48.

4.3.4   Summary

The above treatment of bridge rectifiers has shown how the output waveforms and voltages are obtained for a wide variety of possible operating conditions. Where no overlap interference takes place, the

73

simple method of voltage–time area yields the output voltage with no difficulty. Overlap interference can be an important factor, since, for the single-phase, half-controlled bridge, it demands gate-firing pulses longer than the diode overlap, as well as modifying the waveforms considerably. Similar restrictions occur for the 3-phase bridge with differing delays α_1 and α_2. The control characteristics for bridge rectifiers are conveniently summarised in Fig. 52, which does not, however, include the minor changes in control characteristics resulting from overlap interference.

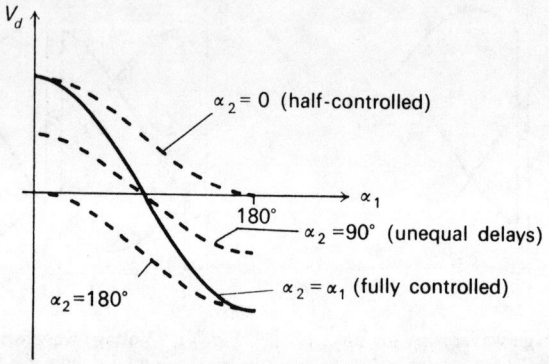

Fig. 52   General control characteristics for 3-phase bridge
Small discontinuities introduced by overlap interferences have been omitted

4.4   Characteristics to short circuit

As for centre-tap circuits, the combination of an excessive output current I_d and a high leakage inductance h gives rise to very prolonged overlaps. Bridge-rectifier behaviour for these conditions is considered for diodes only, as the electronic current controls present for thyristor rectifiers prevent excessive currents from developing. A smooth output current I_d is assumed throughout the operating modes: the short-circuit condition implies an inductive load of zero resistance, for which the conduction pattern can be different from that obtained for a noninductive short circuit.

4.4.1 Single-phase bridge

The single-phase bridge ($q = 2$) has a straight-line regulation characteristic from the no-load voltage to the short-circuit current (Fig. 53). As the overlaps are prolonged, proportionally more voltage–time area is lost from the output waveform, resulting in the linear characteristic. The alternating bridge current has longer periods of

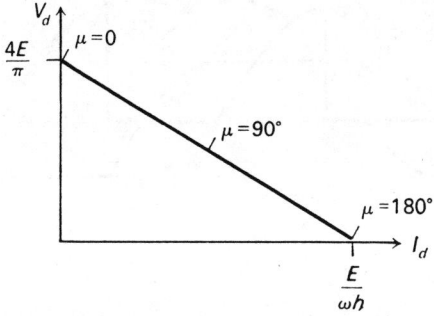

Fig. 53   Regulation characteristic from no load to short circuit for a single-phase bridge

commutation and shorter periods of constant value ($= I_d$) until, at short circuit, the alternating current becomes sinusoidal, as expected for the condition where the rectifier bridge itself also becomes a short circuit. The waveforms are shown for $\mu = 90°$ and $\mu \simeq 180°$ in Figs. 54a and b.

4.4.2 3-phase bridge

The 3-phase bridge is the only polyphase bridge considered, as the 6-phase bridge has no less ripple, nor is it at a higher frequency than the 3-phase bridge. All the advantages lie with the use of two or more 3-phase bridges operating with phase-shifted supplies (Chapter 5).

Whereas, for centre-tap rectifiers, a single rule emerged by which the mode of operation was numerically equal to the minimum number of conducting diodes, such is *not* the case for the 3-phase bridge. The difference lies in the 3-phase bridge's ability to develop overlap

75

interference, where an overlap in the lower half delays the start of an overlap in the upper half. This situation cannot arise for the centre-tap rectifier having only one 'half'.

The behaviour of a 3-phase bridge for increasing overlap angles is nevertheless subdivided into modes which are as clearly, but not as

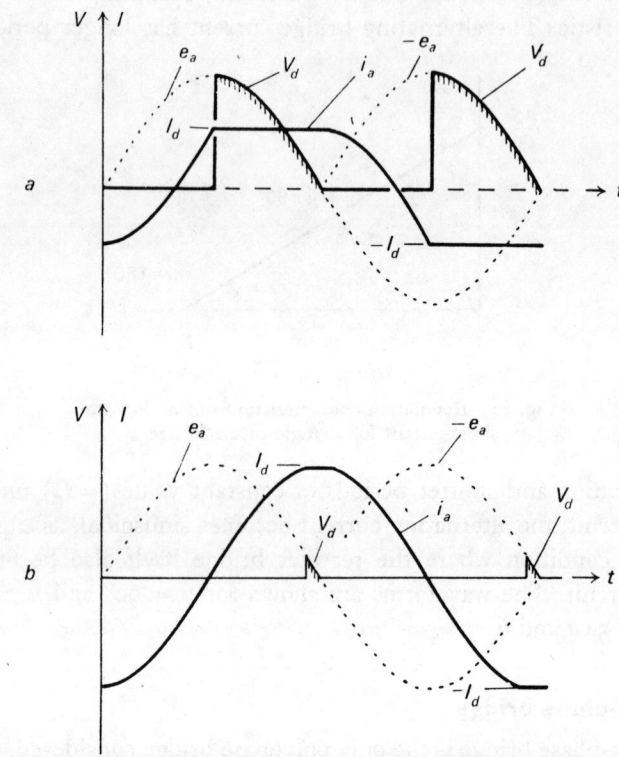

Fig. 54   Output voltage V_d and alternating current i_a for single-phase bridge with total secondary voltage $e_a = E\sin\omega t$

a Overlap = 90°   b Overlap approaching 180°

simply, defined as for centre-tap rectifiers (Hölters, Freris, 1961; Dortort, 1953). For mode 1, where there is no overlap interference, the regulation characteristic is a straight line since, as overlap takes proportionally more voltage–time area, there is proportionally less

76

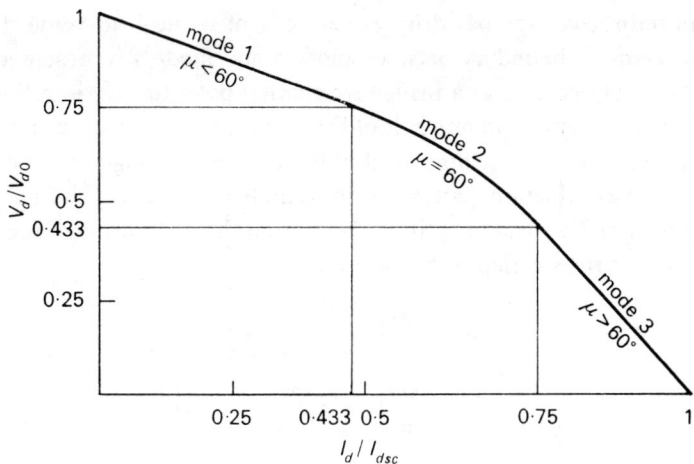

Fig. 55 Regulation characteristic from no load to short circuit for 3-phase bridge, showing modes of operation and overlap angles

$$V_{d0} = 3\sqrt{3}E/\pi \qquad\qquad I_{dsc} = E/\omega h$$

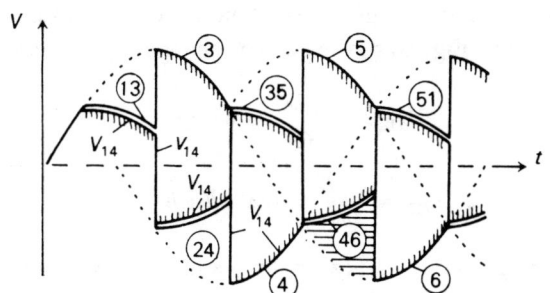

Fig. 56 Voltage waveforms for 3-phase bridge at mode 1–2 boundary
Circled numbers indicate conducting diodes
Horizontally shaded area $= I_d h$

for the output voltage–time area. The overlap interference which occurs throughout mode 2 has the effect of introducing a firing-angle delay which increases with current, deflecting the regulation characteristic from the usual straight line. Mode 3 exhibits the straight-line characteristic like mode 1. The entire regulation characteristic is shown in Fig. 55.

Mode 1 applies from no load up to an overlap of 60°, at which the mean output voltage has dropped to 75% of its no-load value. It is clear from the boundary between mode 1 and mode 2 represented in Fig. 56 that, as soon as a bridge alternating potential V_{14} is released from an overlap μ_{13} on one half of the bridge, it immediately engages in an overlap μ_{24} on the other half of the bridge. Throughout mode 1, the diode conduction pattern is the familiar 2–3–2–3. The output voltage in mode 1 can be written down from the waveforms of Fig. 56 in terms of the overlap μ, as follows:

$$V_{d1} = \frac{3\omega}{\pi}\left(\frac{\sqrt{3}E}{\omega} - I_d h\right) \tag{4.17}$$

$$= \frac{3\omega}{\pi}\left(\frac{\sqrt{3}E}{\omega} - \frac{\sqrt{3}E}{2\omega}2\sin^2\frac{\mu}{2}\right) \tag{4.18}$$

$$= \frac{3\sqrt{3}E}{\pi}\left(\tfrac{1}{2} + \tfrac{1}{2}\cos\mu\right) \tag{4.19}$$

Substituting $\mu = 60°$ yields $V_{d1}/V_{d0} = 0.75$ as stated above.

The current I_{d12} at the modes 1–2 boundary is found by equating the shaded area in Fig. 56 to the current change it produces in h; i.e.

$$I_{d12}h = \frac{\sqrt{3}E}{2\omega}2\sin^2\frac{60°}{2} \tag{4.20}$$

i.e.

$$I_{d12} = \frac{\sqrt{3}E}{4\omega h} = 0.432E/\omega h \tag{4.21}$$

In mode 2, the overlap is always 60°, and there are always three diodes conducting, giving a 3–3–3–3 conduction pattern. At the boundary between modes 1 and 2, it is apparent that, as soon as two bridge a.c. voltages are released from an overlap in one half of the bridge, one of them immediately engages in an overlap in the other half of the bridge. A delay in the end of one overlap automatically delays the start of the next, thus explaining why the overlap remains constant at 60°.

The delay introduced by overlap interference increases with current, having an effect similar to a firing-delay angle, since the conduction of each device is delayed by α_μ. The output is now reduced for two reasons: first, the increase in the voltage–time area spent in changing the current in the leakage inductances, and secondly, the increase of α_μ. The regulation characteristic for mode 2 therefore departs from the straight line and curves downward.

As with centre-tap rectifiers in which the simple definition of overlap angle was found inadequate for higher-mode operation, so also for the 3-phase bridge, where the inadequacy lies in the fact that the overlap does not change with increasing currents in mode 2. A pseudo-overlap $\mu' = \mu + \alpha_\mu$ is similarly defined as the angle between the start of diode-current decay at a light load and diode extinction. This

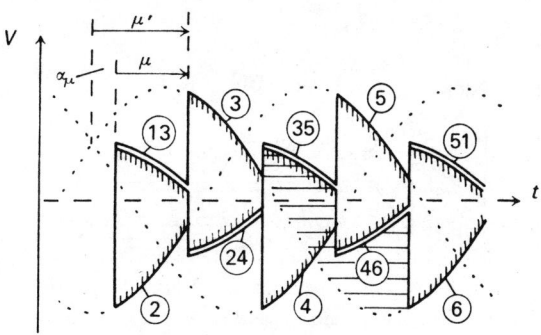

Fig. 57   Voltage waveforms and conducting diodes
for operation of 3-phase bridge in mode 2

Pseudo-overlap angle μ' is introduced, with the overlap-induced delay angle α_μ
Horizontally shaded area changes the current through h_b from $+I_d$ to $-I_d$

definition is used in Fig. 57, which shows the waveforms for the middle of mode 2, from which the output voltage is written down in terms of μ' as

$$V_{d2} = \frac{3\omega}{\pi}\frac{3E}{2\omega}\left(2\sin^2\frac{90-(\mu'-60)}{2} - 2\sin^2\frac{90-\mu'}{2}\right) \quad (4.22)$$

which gives the same result as eqn. 4.19 for $\mu = 60°$. Mode 2 ends when $\mu' = 90°$ ($\alpha_\mu = 30°$); the output voltage V_{d2} is $9E/4\pi = 0.433$ of the no-load voltage, and the instantaneous output voltage momentarily drops to zero for the first time.

The output current I_{d23} at the boundary between modes 2 and 3 is found by equating an area similar to the shaded area on Fig. 57, except that $\mu' = 90°$, with $2I_{d23}h$, since, at one end of this area, the current in the phase b is I_d while, at the other end, it is $-I_d$. The dashed line at zero potential forms a boundary for two obviously equal areas above and below this line, so that

$$2I_{d23}h = \frac{E}{\omega} 2\sin^2\frac{120}{2} \tag{4.23}$$

i.e.

$$I_{d23} = \frac{3E}{4\omega h} = 0 \cdot 75E/\omega h \tag{4.24}$$

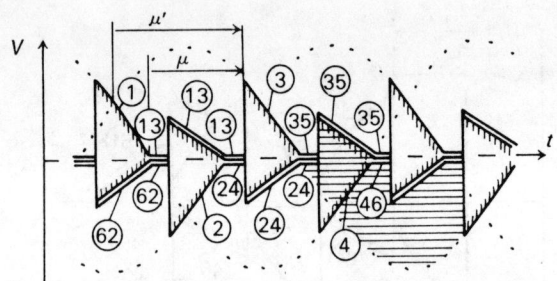

Fig. 58 Voltage waveforms and conducting diodes for operation of 3-phase bridge in mode 3

Note the periods of zero bridge output voltage when four diodes conduct together
Horizontally shaded area changes the current through h_b from $+I_d$ to $-I_d$

Mode-3 operation involves overlaps of greater than 60°, resulting in a conduction pattern 3–4–3–4, which, for four devices conducting together, implies a period of zero bridge-output voltage since a freewheeling path must exist. The overlap increases throughout mode 3, from 60° to 120°, while the pseudo-overlap μ' varies from 90° to 150°, as shown in Fig. 58.

The output voltage in mode 3 is given in terms of the pseudo-overlap μ' as follows:

$$V_{d3} = \frac{3\omega}{\pi}\frac{3}{2}\frac{E}{\omega} 2\sin^2\frac{150-\mu'}{2} \tag{4.25}$$

which gives the same output voltage as eqn. 4.22 at $\mu' = 90°$, and zero at $\mu' = 150°$.

80

The voltage–time area which reduces the phase current through h_b is shaded horizontally where $i_{3'}$ is reduced (or $i_{6'}$ increased). The period of zero output voltage in the middle of this shaded area has diodes $3'$ and $6'$ conducting together, the current in diode $6'$ rising while that in $3'$ falls. It is during this period that the phase current in h_b reverses. The total shaded area brings about a change of current $2I_d$ and hence has an area $2I_d h$. The current i_b ($= i_{3'} - i_{6'}$) approaches the shape of a sine wave, a shape it reaches at the end of mode 3. The short-circuit current I_{dsc} is equal to the peak phase current I_{ap} at short circuit; i.e.

$$I_{dsc} = I_{ap} = E/\omega h \qquad (4.26)$$

This is readily confirmed from the shaded area $2I_d h$ at short circuit, which becomes $2E/\omega$, giving the same result.

The rectifier regulation of a 3-phase bridge through its three modes of operation, shown in Fig. 55, gives the currents and voltages at mode boundaries in terms of the short-circuit current I_{dsc} and the no load voltage V_{d0}.

It is interesting to note the markedly different shape from that obtained for polyphase centre-tap rectifiers, which do not exhibit the delaying effect of overlap interference.

4.5 A.C. ratings

Bridge rectifiers, in contrast to centre-tap, do not require a transformer as an essential part of the rectifier circuit. A transformer may be present for a high-power bridge rectifier where it is appropriate to take the a.c. supply from a 6·6 kV (or high-voltage) distribution system, and where this voltage is too high for the load. Many low-power rectifiers are connected directly to the 415 V 3-phase supply; the rectifier variable-speed drive of up to 500 kW is a notable example.

When considering the current waveforms drawn by bridge rectifiers, the assumption of zero leakage (or supply) inductance is made. The current waveforms are thus rectangular, and it is simple to calculate the r.m.s. value of the alternating line current to the bridge, and hence the apparent power (VA) required from the supply. If a transformer is required, the ratings of primary and secondary are both equal to the apparent-power rating of the bridge. Where there is a connection change between primary and secondary, e.g. star delta, the m.m.f.-balance condition is used to determine the primary-current waveform, which may not be the same shape as the secondary waveform, but, for unity ratio, it will have the same r.m.s. value.

In a fully controlled bridge, a firing angle α delays the entire a.c. waveform by an angle α with respect to the voltage waveform. A power factor of zero corresponds to a firing delay of 90° and zero direct-voltage output.

The alternating-current waveform for a half-controlled bridge differs from that for a fully controlled bridge for delay angles other than $\alpha = 0$. With $\alpha > 0°$ for single-phase and $\alpha > 60°$ for 3-phase, a freewheeling action develops during which the load current I_d does not flow in the a.c. supply lines.

There is no longer a constant relationship between the output current I_d and the r.m.s. alternating current I_{ar} for all firing angles as there was for the fully controlled bridge; but I_{ar}, and hence the apparent power, both decrease with increasing α for constant I_d. The supply system must be designed for the maximum apparent power at zero delay; and, for this condition, the fully controlled and half-controlled bridges are alike.

4.5.1 Single-phase fully controlled bridge

The a.c. waveform for a firing-angle delay α is shown in Fig. 59 in relation to the sinusoidal supply voltage. For zero delay, i.e. maximum output, the output power P_d is

$$P_d = \frac{4EI_d}{\pi} \qquad (4.27)$$

The apparent-power input is given by

$$VA_r = \sqrt{2}\ EI_d \qquad (4.28)$$

Hence $\qquad P_d : VA_r = 1 : \dfrac{\sqrt{2}\,\pi}{4} = 1 : 1 \cdot 11 \qquad (4.29)$

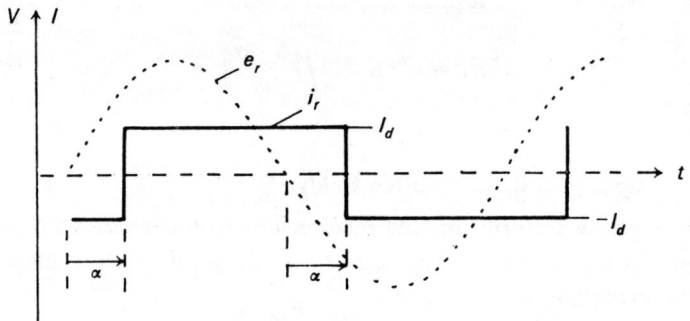

Fig. 59   Alternating current for fully controlled single-phase bridge
operating with delay α (neglecting leakage inductance)

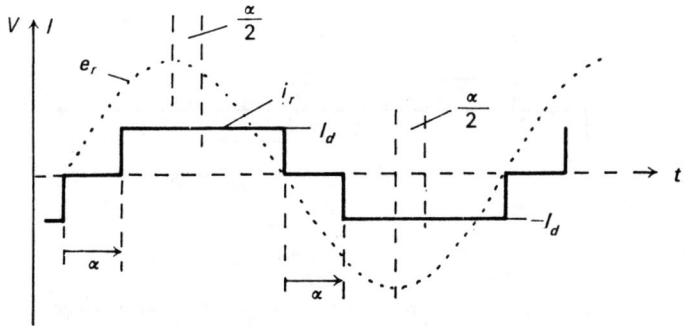

Fig. 60   Alternating current for half-controlled single-phase
bridge operating with delay α (neglecting leakage inductance)
Freewheeling periods coincide with zero alternating current

4.5.2   Single-phase half-controlled bridge

The a.c. waveform for a firing delay α is shown in Fig. 60. For zero
delay, the results are the same as for the fully controlled bridge, and
eqns. 4.27, 4.28 and 4.29 apply. For a delay α, the output power P_d is

$$P_d = \frac{4E}{\pi} \left(\tfrac{1}{2} + \tfrac{1}{2}\cos\alpha\right) I_d \qquad (4.30)$$

The apparent-power input is

$$VA_r = \sqrt{2}\,E\,I_d\sqrt{\left(\frac{\pi-\alpha}{\pi}\right)} \qquad (4.31)$$

4.5.3 3-phase fully controlled bridge

The a.c. waveform and its relationship to the phase voltage are shown in Fig. 61 for a delay α. For zero delay, the output power P_d of the bridge is

$$P_d = \frac{3\sqrt{3}EI_d}{\pi} \qquad (4.32)$$

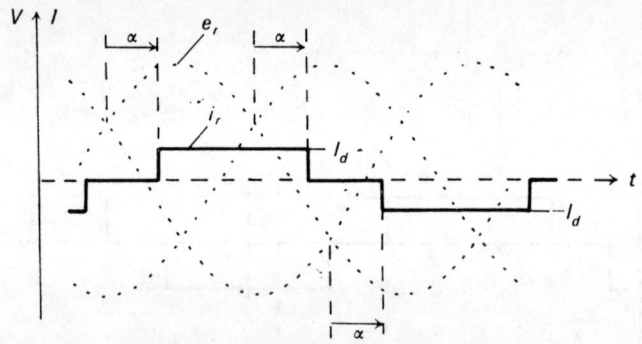

Fig. 61   Alternating current for 3-phase fully controlled bridge operating with delay α (neglecting leakage inductance)

The apparent-power input for three phases is

$$VA_{rst} = 3\,\frac{E}{\sqrt{2}}\,I_d\sqrt{(\tfrac{2}{3})} \qquad (4.33)$$

Hence $\qquad P_d : VA_{rst} = \dfrac{3\sqrt{3}}{\pi} : \sqrt{3} = 1 : 1{\cdot}047 \qquad (4.34)$

Equation 4.34 shows how little extra apparent power is required for the 3-phase bridge compared with its power rating at zero delay.

4.5.4 3-phase half-controlled bridge

The a.c. waveform for a delay α is shown in Fig. 62. The positive halfcycle carried by a thyristor is delayed by α while the negative halfcycle carried by a diode remains in the same position for all α. For $\alpha < 60°$, both halfcycles are of 120° duration, as shown in the upper current waveform of Fig. 62a. But, for $\alpha > 60°$, the halfcycle flowing through the thyristor continues beyond the start of diode

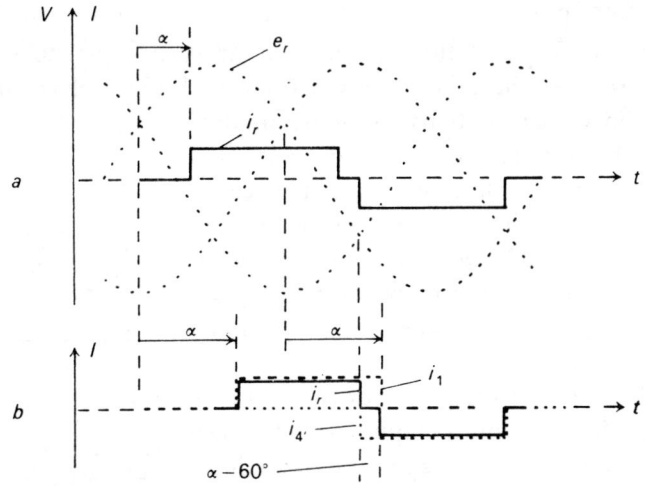

Fig. 62   Alternating current for 3-phase half-controlled
bridge operating with delay α

(a) $\alpha < 60°$
(b) $\alpha > 60°$

Note that, for (b), freewheeling occurs as devices 1 and 4' conduct together for $\alpha - 60°$

conduction. If leakage inductance were not assumed to be zero, over-lap interference would occur as described in Section 4.2.2. For zero leakage inductance, freewheeling starts when $\alpha > 60°$, and lasts for $\alpha - 60°$, during which time the alternating current is zero. Fig. 62b shows the current waveform for $\alpha \simeq 78°$, for which a freewheeling period of 18° exists while the thyristor and the diode connected to the same phase conduct together.

For zero delay, the output power and apparent-power input are the same as for a fully controlled bridge (eqns. 4.32, 4.33 and 4.34).

For firing delays less than 60°, the duration of the positive block of current is constant, but its timing is delayed by α until, at $\alpha > 60°$, cancellation occurs between the trailing edge of the thyristor curve and the leading edge of the diode current, giving a freewheeling period of $\alpha - 60°$.

4.5.5 Summary

The waveforms of the line currents (Wallach, 1965) to various bridge rectifiers have been derived from the bridge behaviour discussed in earlier Sections, using the simplifying assumption of zero leakage inductance. The r.m.s. value of the line current, and hence the apparent-power input to the bridge from the supply, are readily calculated. A firing angle $\alpha > 0$ delays the current waveform with respect to the voltage, contributing to a lower power factor.

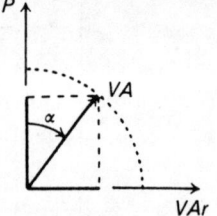

Fully controlled bridges have a constant ratio of r.m.s. line current to output current I_d, and hence present the same apparent-power loading for all delays when I_d is constant. The power factor becomes zero (see Fig. 63) when a firing-angle delay of 90° gives zero direct output voltage.

Fig. 63 Power locus for fully controlled rectifier showing constant apparent power (VA), decreasing power and increasing reactive power (VAr) as the delay α increases

Half-controlled rectifiers develop a freewheeling action at a certain firing-angle delay ($\alpha = 0°$ for single-phase, 60° for 3-phase) beyond which the r.m.s. line current decreases with increasing α, and the power factor also decreases.

A further discussion of power factor and a treatment of harmonic currents is given in Chapter 5, where principles common to all forms of rectifier are presented.

4.6 Reactance and regulation

Although bridge rectifiers operate frequently without a transformer, it is common for some reactance to be inserted between the supply and the bridge, where the impedance of the supply is low, to limit fault current.

The voltage drop of a rectifier under load, according to the analysis of this Chapter, is governed by the effective inductance in series with the a.c. supply, be it transformer leakage inductance (Feinberg and Chen, 1964b), added inductance or the inductance of the supply itself. The reactance of the inductance, particularly when it is leakage reactance of a transformer, is often referred to as a percentage of the rated transformer impedance. It is convenient to relate this percentage reactance to the percentage voltage drop of rectifier output which it produces. The percentage reactance defined as the percentage of the rated voltage required to drive the rated r.m.s. current through the reactance is related to the rectifier behaviour through the r.m.s. value of the nonsinusoidal rectifier currents.

The single-phase diode bridge has an r.m.s. alternating current equal to the output current I_d, which gives, for a percentage reactance x,

$$\frac{x}{100} \frac{2E}{\sqrt{2}} = I_{dx}\,\omega 2h \tag{4.35}$$

On combining this with the equation for the output voltage of a single-phase bridge, i.e.

$$V_d = \frac{4E}{\pi} - \frac{4I_d\omega h}{\pi} \tag{4.36}$$

we have

$$\frac{100 V_d}{4E/\pi} = 100 - \frac{x}{\sqrt{2}} \tag{4.37}$$

The 3-phase bridge has an r.m.s. value of alternating current equal to $I_d\sqrt{2}/\sqrt{3}$. Considering one phase only, we have

$$\frac{x}{100} \frac{E}{\sqrt{2}} = I_d\omega h\sqrt{2}/\sqrt{3} \tag{4.38}$$

which, when combined with the voltage equation

$$V_d = \frac{3\sqrt{3}E}{\pi} - \frac{3I_d\omega h}{\pi} \qquad (4.39)$$

gives

$$\frac{100V_d}{3\sqrt{3}E/\pi} = 100 - \frac{x}{2} \qquad (4.40)$$

The 3-phase bridge rectifier thus has a smaller regulation than the single-phase bridge. Where a phase delay reduces the output of a controlled rectifier, the actual voltage drop is the same as for a diode rectifier, but it will be a greater percentage of the reduced no-load rectifier output.

5 RECTIFIER APPLICATIONS, SUPPLIES AND LOADS

5.1 Introduction

Rectifier supply currents, their power factor and harmonic content are of fundamental importance to the correct design of the a.c. supply system to the rectifier and to the correct operation of the remainder of the supply network with the rectifier connected. The rectifier differs from most other a.c. plant, in that it draws a nonsinusoidal current having a rectangular or stepped waveform. Assuming a zero-impedance supply, the harmonic currents do not produce harmonic voltage drops; in other words, the supply waveform remains sinusoidal. The only component of current which can consume power is thus the fundamental. The presence of harmonics (Evans and Muller, 1939), however, places an extra current load on the system, since the r.m.s. value of the actual current is greater than the r.m.s. value of its fundamental component. In the calculation of supply current, the efficiency of a silicon rectifier and its transformer, where present, are assumed to be 100%. The input power and output power are thus assumed equal, as in Sections 3.5 and 4.4. It is the fundamental component of current and its phase relationship to the voltage which determines the input power.

The supply system meets the apparent-power (VA) demand of the rectifier, determined by the r.m.s. value of the rectifier current. It is thus required to determine the apparent-power input to a rectifier and its power factor (Schmidt, 1958) from the d.c. output parameters, and to consider the presence and effects of harmonic currents. The apparent-power input to various rectifier circuits has been calculated in Sections 3.5 and 4.4, but the concept of power factor and the effects of harmonics were not developed. Where the rectifier rating is a large proportion of the system rating at the point of connection of the rectifier, called the point of common coupling, the assumption

89

of zero supply-system impedance is invalid. Voltage-waveform disturbances occur (Gerecke, 1960), particularly at commutations where the rates of change of current are highest. The presence of harmonic voltages on the system can cause excessive harmonic currents to flow even in remote power-factor-correcting capacitors (Brownsey and Csuros, 1963) and perhaps instability (Ainsworth, 1967). The harmonic currents always present can cause maloperation of relay and protective gear, and can induce high-frequency voltages in adjacent circuits, of particular importance when these are sensitive communication or telemetry circuits (Von Zastrow).

To protect other consumers, the UK Electricity Council has prepared recommendations (G5/2, 1967) for the allowable proportion of harmonic currents on a supply system to which rectifiers are connected. Small convertors (100 kW) rarely produce excessive harmonic currents, provided that the total rating is small compared with the supply rating. Large convertors for mine winders, steel-rolling mills and, of course, for h.v. d.c. do not satisfy this condition, and it is usually necessary to connect harmonic filters [Cory (Ed.), 1965; Adamson and Hingorani, 1960] at the a.c. terminals of the convertor to bypass harmonic currents, i.e. to lower the reactance of the supply to harmonics. H.V. d.c. convertors always use 12-(or more)-pulse operation ($p \geqslant 12$), and embody complex harmonic filters which represent a significant proportion of the cost, having a reactive-power rating approximately half the power rating of the convertor itself. These capacitors fulfil a dual role of harmonic filtering and power-factor correction.

For a constant output current I_d, the apparent-power demand of a fully controlled rectifier remains constant in spite of a reduction of output voltage and power by a firing-angle delay. As the power demand of a rectifier falls with its power output, it is apparent that the demand for reactive power must increase with an increasing delay α (Fig. 63). Prolonged operation of a controlled rectifier at less than the maximum output consequently applies a heavy lagging reactive load on the system, which may for economic reasons require power-factor correction. The capacitors for this purpose will require series inductance to limit the harmonic current which results if the supply does not have zero reactance.

The above introductory comments are sufficient to emphasise the importance of nonsinusoidal rectifier currents and the heavy reactive-power requirements of a controlled rectifier operating at reduced output voltage. One of the aims of multiple rectifier arrangements, using two or more basic circuits, is to mitigate the obvious problems of harmonic currents and poor power factor at reduced voltage. The combination of basic rectifier circuits is described in Section 5.4.

While the analysis of rectifier behaviour has been presented in Chapters 3 and 4 for smooth output current, implying a load of infinite inductance, practical applications do not satisfy this condition. However, for rectifiers of high pulse number p, even supplying a purely resistive load, the current ripple is small, and hence the waveforms do not depart significantly from those for perfectly smooth current. Rectifiers with a low pulse number, e.g. centre-tap circuits with $q = 2$ or 3 and the single-phase bridge ($q = 2$) have an output-current ripple which may not be negligible. The general behaviour of rectifiers with other than highly inductive loads is discussed in Section 5.5.

All semiconductor rectifiers, whether supplying rotating machines, works d.c. systems or large electrolytic loads, require protection against overvoltage and overcurrent. The principles of rectifier protection are presented in Section 5.6. The detailed protective circuits are likely to vary with the particular application, but the principles described underlie most protective systems.

5.2 Power factor

The nonsinusoidal waveform of current drawn by a rectifier requires that power factor be redefined as

$$\text{total power factor} = \frac{\text{power output, W}}{\text{apparent-power input, VA}} \qquad (5.1)$$

It has already been shown that, even with zero leakage inductance, giving rectangular current waveforms, which for zero delay are in phase with the voltage, rectifiers consume more apparent power than they produce power output. The power factor of a rectifier circuit thus includes a term describing the effect of the nonsinusoidal current. The term, called the 'distortion factor', is the ratio

$$\text{distortion factor } v = \frac{\text{r.m.s. value of the fundamental current}}{\text{r.m.s. value of the total current}}$$

(5.2)

Only the fundamental consumes power from a supply of zero reactance (for which the supply voltage remains sinusoidal) despite the presence of harmonic currents. Eqns. 3.44, 3.48, 3.49, 4.29 and 4.34, calculated for maximum output, yield the corresponding distortion factors of the various circuits.

Table 2

Rectifier circuit ($h = 0$)	Distortion factor v		
	$q = 2$	$q = 3$	$q = 6$
Centre-tap	0·90	0·83	0·83
Fully controlled bridge	0·90	0·955	—

The distortion factors of Table 2 are governed solely by the rectifier current waveform, and not by any phase-angle relationship it or its fundamental may have to the voltage waveform.

In reality, the current waveforms, particularly of diode rectifiers, depart noticeably from the rectangular shape assumed so far, owing to overlap, which for diodes is not negligible. (Controlled rectifiers operating at a substantial delay have a much shorter overlap angle for which the rectangular assumption is acceptable.) Overlap has two opposite effects on the total power factor: by sloping the rectangular edges of the current waveform, the distortion factor is increased towards unity; by delaying both the rise and the fall of current, the fundamental component of the current is phase-delayed with respect to the voltage (Uhlmann, 1955). The phase shift ϕ of the fundamental current with respect to the voltage determines the 'displacement factor':

$$\text{displacement factor} = \cos\phi$$

(5.3)

A firing-angle delay α contributes directly to the phase shift ϕ; for a fully controlled rectifier with zero overlap, $\phi = \alpha$; but, for finite overlap, $\phi > \alpha$.

The power factor of a rectifier is the product of the distortion and displacement factors:

$$\text{total power factor }(\cos\phi_1) = \text{displacement factor} \times \text{distortion factor }(v)$$
$$(5.4)$$

For rectifiers with a pulse number of six or more, IEC and BSI declare that, for practical reasons, the total power factor $\cos\phi_1$ should be taken as the displacement factor $\cos\phi$ (BS 4417, 1969; IEC 146, 1963).

Generally, the lower the pulse number of a rectifier, the farther from sinusoidal is its a.c. waveform, the lower its distortion factor and the greater the harmonic content of the waveform. The exception to this generality appears on a comparison of the centre-tap circuits in Table 2 for $q = 2$ and $q = 3$; the latter, owing to the differing shapes of positive and negative halfcycles of primary current, has a lower distortion factor (i.e. more distortion) than the $q = 2$ circuit. Higher distortion factors can be obtained by adding the currents of several rectifiers operating from supplies of differing phase; this increases the pulse number and gives a supply-current waveform with more steps per cycle. The case of two 3-phase bridges is represented in Fig. 64 where star–delta secondaries provide the 30° phase shift required. With series connection of bridge outputs, I_d is common to both bridges.

The displacement factor $\cos\phi$ is governed primarily by the firing-angle delay α, although overlap μ also contributes where α is small.

Fully controlled rectifiers (i.e. fully controlled bridges and centre-tap rectifiers without freewheeling diodes) have a displacement factor $\cos\phi$ which can never be greater than $\cos\alpha$. The power factor of such a rectifier is inevitably low at low output voltage. The half-controlled bridge, or any rectifier with a freewheeling diode, offers some improvement since, although the power factor is zero at zero output voltage, the apparent power is also zero, since the bridge is freewheeling virtually all the time, and the alternating currents are thus of very short duration. Fig. 65 illustrates the variation of apparent power and displacement factor for half- and fully controlled rectifiers.

93

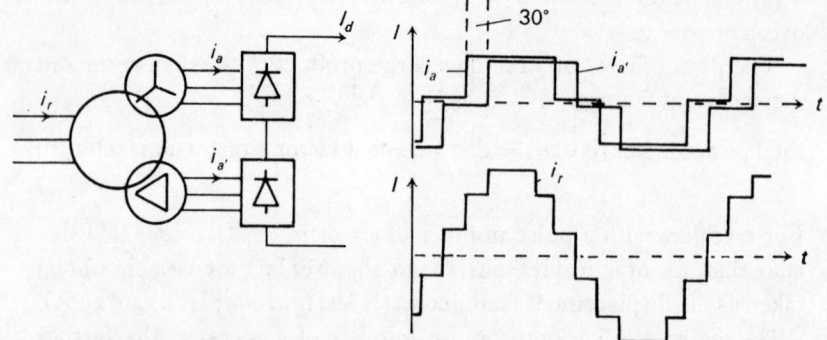

Fig. 64 Improvement in shape of primary-current waveform when two 3-phase bridges share the load and are supplied with alternating voltages with a relative phase shift of 30°

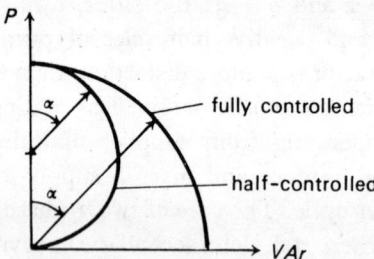

Fig. 65 Reduced apparent-power demand of a half-controlled rectifier when operating at considerably reduced output voltage ($\alpha \gg 0$)

The same effect as a half-controlled bridge is obtained when a fully controlled bridge and a diode bridge of the same maximum voltage are series-connected, and their a.c. terminals are supplied from two transformer secondaries of the same phase. For low output voltages, the fully controlled bridge operates as an invertor, cancelling most of the rectifier current waveform (Fig. 66).

A substantial improvement in power factor can be obtained by controlling series-connected rectifier circuits sequentially, rather than applying the same firing-angle delay to all circuits. The rectifier circuits must possess a freewheeling path to allow the current output of one circuit to bypass the nonoperative circuits. A good example

94

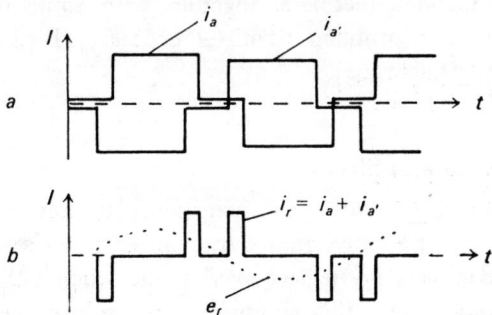

Fig. 66 (*a*) Current waveforms of two series-connected 3-phase bridges, one uncontrolled and the other operating with $\alpha = 160°$ (*b*) Primary current of transformer supplying the two bridges, showing large degree of cancellation of secondary currents

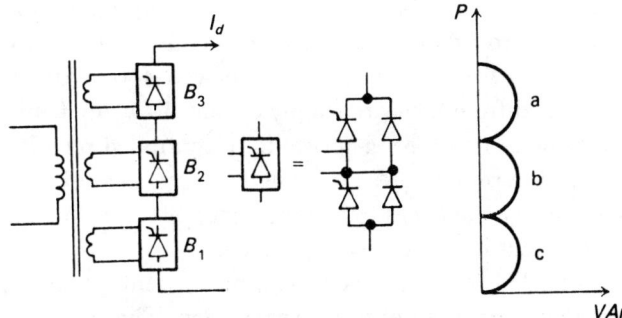

Fig. 67 Substantial reduction of reactive power when sequential control of series-connected rectifiers is adopted

$$a \begin{cases} V_{B1} = V_{B2} = \text{maximum} \\ V_{B3} = \text{controlled} \end{cases} \qquad b \begin{cases} V_{B1} = \text{maximum} \\ V_{B2} = \text{controlled} \\ V_{B3} = 0 \end{cases}$$

$$c \begin{cases} V_{B1} \text{ controlled} \\ V_{B2} = V_{B3} = 0 \end{cases}$$

of this technique is the use of several asymmetric single-phase half-controlled bridges to provide variable output voltage for the traction motors of a single-phase locomotive (Onoda *et al.* 1969). Fig. 67 shows the arrangement and the associated relationship between power and reactive power.

Numerous variations on the above techniques have been applied to improve the power factor and/or lower the harmonic content of current of controlled rectifiers, together with some more drastic approaches to the problem (Toth *et al.* 1963; Bird *et al.* 1969; Hingorani and Hall, 1965).

5.3 Harmonics and filters

The harmonics present in the nonsinusoidal alternating currents drawn by a rectifier are determined by Fourier analysis of the current waveform. It is usual to neglect overlap and work with rectangular or stepped waveforms; this simplifying procedure overemphasises the higher harmonics present, but these are easiest to filter.

Current waveforms having the same shape for positive and negative halfcycles and symmetry about the centre of the halfcycle do not possess any even harmonics. The 3-phase centre-tap rectifier and the 3-phase half-controlled bridge are the only rectifiers considered which do possess even harmonics in their current waveform. Any 3-phase rectifier supplied from a 3-wire supply cannot draw any triplen harmonics. A general rule for a 3-phase, fully controlled rectifier states that the lowest harmonic present is one less than the pulse number. A 3-phase bridge thus has 5th, 7th, 11th, 13th, 17th, 19th etc. harmonics present in its alternating current.

Fully controlled rectifiers draw the same current waveform from the a.c. supply, regardless of firing-angle delay. The relative magnitudes of the harmonics present are thus unchanged: their phases with respect to the voltage do change; and, in particular, the fundamental current lags the voltage by the displacement angle ϕ, which, for zero overlap, is equal to the firing delay α.

Half-controlled rectifiers, or rectifiers with a freewheeling diode, draw a current waveform in which, for firing delays greater than a certain value, the duration of current flow in each halfcycle becomes shorter. There is thus a change of current waveform and a consequent change in the harmonic content.

Fig. 68 shows the magnitudes of the harmonics present in various rectifier supply currents, and how they vary with firing angle α.

Harmonic filters are not usually required until the power has reached a level which makes the connection of two or more rectifier

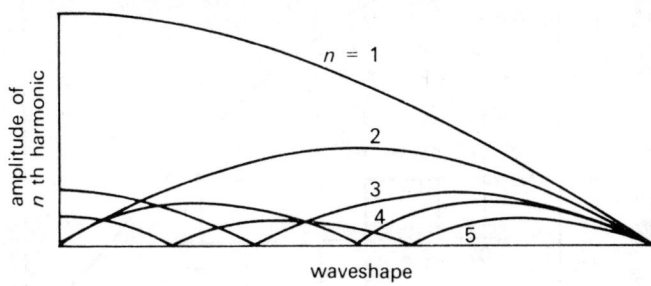

Fig. 68A

Conventional illustration of harmonic amplitudes as a function of waveshape. Figs. 68c and D are based on this illustration, but they allow one or two curves only to be used for all harmonics, simplifying the diagram and increasing the accuracy. This simplification is achieved by using a horizontal scale for $n\gamma$ or $n\alpha$, which reads in alternating directions for increasing angle

Type of rectifier	Fully controlled	Fully controlled + freewheel diode	Half-controlled
Single-phase centre tap and bridge	$\gamma = 0$	$\gamma = \alpha$	$\gamma = \alpha$
3 – phase bridge	$\gamma = 60°$	$\alpha \leqslant 60°,\ \gamma = 60°$ $\alpha > 60°,\ \gamma = \alpha$	$\gamma = \alpha$ Fig 68D

Fig. 68B

Table giving values of γ for Fig. 68c and α for Fig. 68D for various rectifier circuits operating with delay angle α, assuming the overlap angle is zero

An example of the use of Figs. 68c and D is given below for Fig. 68c: find the 17th harmonic when $\gamma = 60°$:

$$\frac{I_{17.0}}{I_d} = \frac{400}{17\pi}\% = 7\cdot48\% \ (= \text{17th harmonic in square wave})$$

$$17\gamma = 5\tfrac{2}{3}\pi, \quad \frac{I_{17.\,\pi/3}}{I_{17.0}} = 86\cdot6\% \ (\text{from Fig. 68c})$$

$$\frac{I_{17.\,\pi/3}}{I_d} = 6\cdot48\% \quad \text{and} \quad \frac{I_{17.\,\pi/3}}{I_{1.\,\pi/3}} = 5\cdot88\%$$

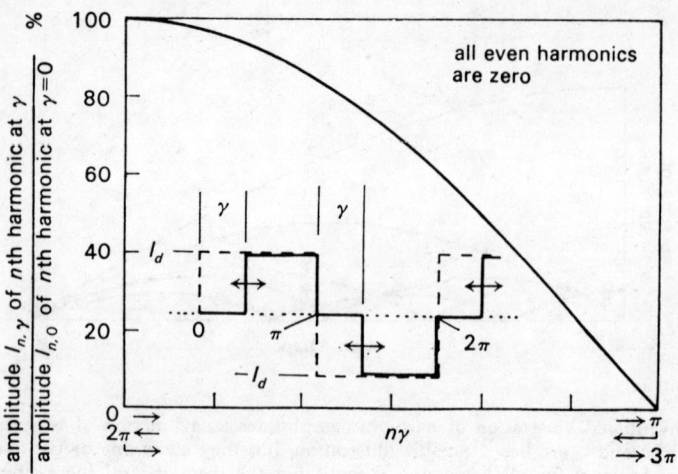

Fig. 68c
Amplitude of the nth odd harmonic $I_{n,\gamma}$ of stepped a.c. waveform shown, expressed as percentage of its amplitude $I_{n,0}$, when $\gamma = 0$. $I_{n,0}$ is obtained from smooth direct-current I_d using the relationship

$$\frac{I_{n,0}}{I_d} = \frac{400}{n\pi} \%$$

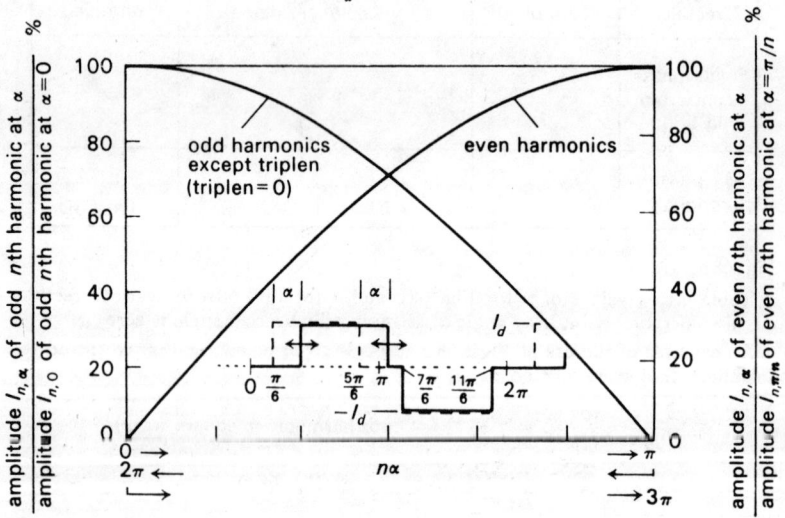

Fig. 68d
Amplitude of nth odd and even harmonics $I_{n,\alpha}$, of the stepped a.c. waveform shown (which corresponds to the alternating supply current of a half-controlled 3-phase bridge). $I_{n,\alpha}$ is expressed as a percentage of $I_{n,0}$ (odd harmonics) or of $I_{n,\pi/n}$ (even harmonics), which are obtained from the direct current I_d using the relationships

$$\frac{I_{n,0}}{I_d} = \frac{200\sqrt{3}}{n\pi} \% \quad \text{(odd harmonics)}$$

$$\frac{I_{n,\pi/n}}{I_d} = \frac{200\sqrt{3}}{n\pi} \% \quad \text{(even harmonics)}$$

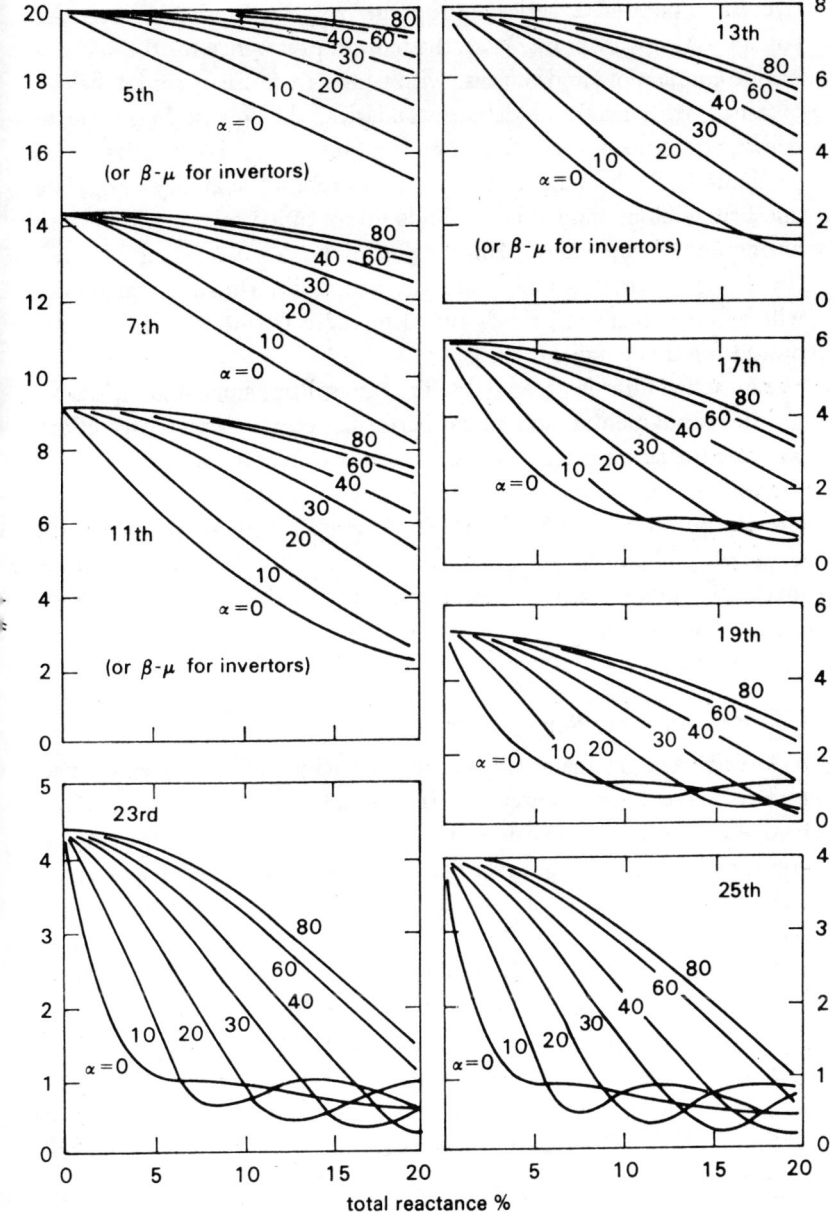

Fig. 68E

Effect of supply inductance which introduces overlap is shown for 3-phase fully controlled bridge. Note that, where α is small, large overlaps provide substantial attenuation of the higher harmonics, but less as α increases when the overlap angle is consequently reduced (Courtesy IEEE) circuits in parallel or series appropriate from the thyristor-current or voltage-rating viewpoint. With the natural preference for the 3-phase bridge connection, and the ease with which 12-pulse operation can be obtained with star and delta secondaries, the lowest harmonic to which a filter need be tuned is the 11th. However, it is often desirable, particularly for h.v. d.c. transmission, to retain the ability to operate one bridge alone, and this demands filters tuned to the 5th and 7th harmonics. Also, where firing-angle differences can occur between two bridges, cancellation of the 5th, 7th, 17th, 19th etc. harmonics will be imperfect, and filters tuned to these frequencies will be required for this condition also.

The shunt filter is preferred to the series filter, since it is capacitive to the fundamental and thus performs power-factor correction, whereas the series filter is inductive to the fundamental and degrades the voltage regulation. The filter for each harmonic is a series *RLC* circuit, but it is usually desirable to combine these in pairs into double-tuned filters, on the grounds that only one capacitor need meet the voltage-surge requirements, and also that resistive losses can be minimised.

5.4 Multiple rectifier circuits

Already several instances of the advantages of multiple rectifier circuits have been mentioned. Individual rectifier circuits, either bridge or centre-tap, can have their d.c. outputs series- or parallel-connected where a single rectifier cannot supply sufficient power or where the harmonic input currents or output ripple voltage of a single rectifier are too great.

The series connection of rectifiers presents no problem. The output voltage is the sum of the voltages of the individual rectifiers, while the output current I_d is common to all. Ideally, the ripple voltages of the rectifiers should be equally phase-displaced from each other, so that the total ripple frequency is multiplied by the number of circuits contributing. Such an arrangement implies that the alternating voltages supplying each rectifier, and hence their alternating currents, are similarly phase-shifted with respect to each other. The supply-current waveform will have more steps per cycle, and hence more of the lower harmonic currents will be absent.

Series-connected 3-phase bridges are universally used for h.v. d.c. transmission, the commonest arrangement being two bridges. Series-connected single-phase bridges have been used for traction-motor voltage control (Onoda *et al.* 1969), according to Fig. 67. More recently, an unusual form of series connection of 3-phase bridges has been proposed, where the intermediate connecting in common of

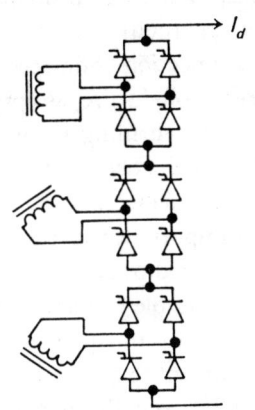

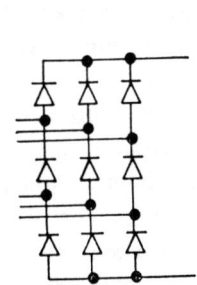

Fig. 69   Dual-input bridge rectifier in which common-ing to form intermediate d.c. connection has been omitted

Fig. 70   Three fully controlled single-phase bridges allow in-clusion of freewheeling period for each bridge in turn, even when inverting, whereby vol-tage control brings a less severe penalty of poor power factor

devices to form an intermediate d.c. terminal has been omitted (Yair *et al.* 1969). Fig. 69 shows the circuit. Little advantage appears to be offered by this arrangement over the series connection of appro-priately chosen conventional 3-phase bridges.

Another recent proposal (Gardner and Fairmaner, 1968), shown in Fig. 70, is for the series connection of three single-phase fully controlled bridges supplied from the secondaries of a 3-phase trans-former. When this arrangement is used as an h.v. d.c. convertor, a special firing-control system allows each bridge in turn to freewheel the direct current I_d twice per cycle. This form of control allows operation at a reduced direct voltage without the severe penalty of a poor power factor.

The increase of thyristor voltage ratings in recent years has meant that, for most industrial applications, a single thyristor has a voltage rating sufficient for applications using the 240 V or 415 V supply. The series connection of devices (Hey; Hall, 1969) is now only necessary for traction thyristor equipment and special high-voltage applications, including, in the near future, h.v. d.c. convertors (*IEE Conf. Publ.* 53, 1969).

Parallel connection of devices or convertors appears to be necessary for the foreseeable future, as applications requiring currents in excess of 100 kA are becoming common (Mori, 1969; Juri and Yoshida, 1969). While such currents are met in part by the parallel connection of the devices themselves, the parallel operation of rectifier circuits is of equal importance, particularly when phase shift between convertors is required to reduce the harmonic currents.

It is not advisable to parallel two rectifiers directly if their output ripple voltages are unequal in magnitude or phase; the result is for the output current to switch rapidly from rectifier to rectifier, flowing predominantly from the one with the higher instantaneous output voltage. Two 3-phase centre-tap circuits so paralleled behave as a diametrical 6-phase star for which the device conduction angles are reduced from 120° to 60°, giving poorer device and transformer utilisation and poorer regulation. To preserve the correct operation of each rectifier, they are paralleled through an interphase transformer (i.p.t.) unless the busbar reactance and/or secondary reactance is sufficient to ensure 120° conduction. It is not unusual to find that the inductance of the busbar system is sufficiently high to avoid the necessity for an i.p.t. Historically, the interphase transformer has always been used with multiple centre-tap rectifiers; now that the bridge rectifier has largely superseded the centre tap, the interphase transformer performs the same role between two bridge rectifiers whose ripple voltages are not in phase (Duff and Ludbrook, 1968). Where four rectifiers are paralleled, three interphase transformers are sometimes used (Fig. 71).

The behaviour of the interphase transformer is described with reference to two 3-phase centre-tap rectifiers with their ripple voltages displaced by 60°. Their output voltage waveforms, shown single-dotted and double-dotted in Fig. 72, are for light-load conditions (Fig. 72a) and full-load conditions (Fig. 72b). Provided that each

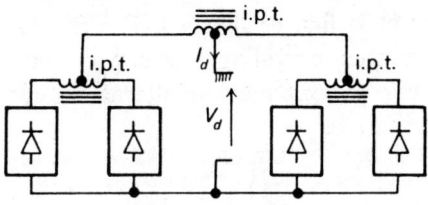

Fig. 71 Use of interphase transformers when paralleling the outputs of four rectifiers whose ripple voltages are phase-displaced to give an improved primary-current waveform

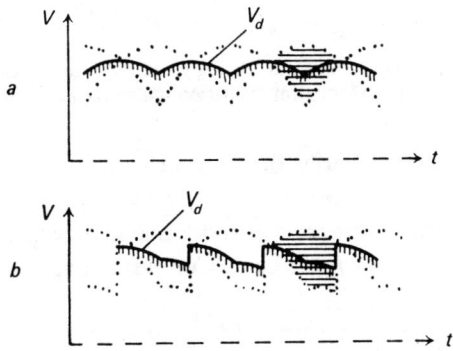

Fig. 72 Output voltage at (a) light load and (b) full load for two 3-phase centre-tap rectifiers paralleled with interphase transformer

Boundary-shaded line is output voltage for continuous output current. Shaded voltage–time area (b) must be supported by i.p.t. without saturation

rectifier operates with continuous current, the potential of the i.p.t. centre tap follows the mean of the two rectifier voltages, shown solid in Fig. 72. It is immediately apparent that the ripple frequency of the output is double that of either rectifier. The voltage-regulation characteristic will be the same as for a single rectifier; but, for two centre-tap rectifiers, the benefit of cancelling residual m.m.f.s is also obtained. In Fig. 72b, the effect of overlap has been included in the output waveform of each rectifier, an effect which increases the voltage–time area of each halfcycle of alternating voltage appearing across the i.p.t. winding.

The inductance of the i.p.t. is calculated from this voltage–time area (shaded) and from the allowable current change produced by it. At a low value of output current I_{dt}, called the transition current, continuous current is first established. For this condition, one can consider each rectifier providing a steady current $I_{dt}/2$, while the shaded area in Fig. 72a produces an alternating current I_{Mp} of peak

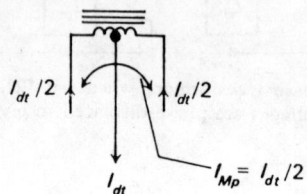

Fig. 73   Direct and alternating currents in interphase
transformer at transition current I_{dt}

magnitude $I_{dt}/2$, which varies the current contribution of each rectifier from zero to I_d and vice versa, as shown in Fig. 73. If the rectifiers must operate with continuous current down to n per cent. of the rated current, $I_{dt} = I_d \times n$ per cent. Consequently,

$$H_{i.p.t.} = \frac{\text{shaded } VTA \text{ (Fig. 72)}}{I_d \times n \text{ per cent.}} \qquad (5.5)$$

For output currents less than the transition current, each rectifier supplies discontinuous current. In Fig. 74, each rectifier output voltage can be considered as a phase voltage, and each half of the i.p.t. as its associated leakage inductance (noting that mutual inductance is present between the halves). One rectifier carries the output current alone until the instantaneous output potential of the other is greater, when current from the incoming rectifier rises while that of the outgoing rectifier falls. During rectifier overlap, the output of the multiple rectifier follows the mean output voltage of the two individual rectifiers until the current carried by the outgoing rectifier reaches zero. At this instant, the output voltage of the multiple rectifier jumps to that of the incoming rectifier.

As the transition current is usually small compared with the rated current, the waveforms of the individual rectifiers at the transition current approximate closely to the no-load waveforms. At full load, however, the shaded area in Fig. 72*b* is noticeably greater owing to overlap, and the i.p.t. must stand this larger voltage–time area without saturation. Note that the individual rectifier output currents produce cancelling m.m.f.s in the i.p.t., so that, with well balanced rectifiers, an air gap is not required.

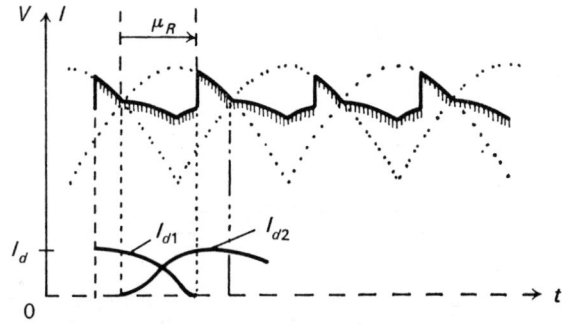

Fig. 74   Output voltage and individual rectifier output current
for operation below transition current

Overlap μ_R occurs while output current I_{d1} of one rectifier
decreases while that of the other (I_{d2}) increases

The whole of the above description has been presented for diode rectifiers. Where the individual rectifiers are controlled, their output voltage waveforms are modified by the firing delay α, which greatly increases the ripple component. The voltage–time area to be supported by the i.p.t. is much greater (Fig. 75), and hence the i.p.t. inductance must also be substantially increased if the transition current is to remain at the same percentage of the rated current. A jump phenomenon (McTaggart, 1968) may also appear.

12-pulse operation is now commonly achieved by connecting two 3-phase bridges in series, fed from star and delta secondaries of the rectifier transformer. Parallel connection using an interphase transformer is more appropriate where the load voltage does not warrant the voltage-doubling effect of the series connection.

18-pulse operation requires three transformers with phase-shifting primaries or one transformer with three phase-shifted secondaries, as shown in Figs. 76a and 76b. Each secondary supplies a bridge rectifier. For low-voltage applications, centre-tap rectifiers paralleled with interphase transformers would be used.

24-pulse operation requires two transformers, each providing 12-pulse operation, with primaries arranged for a 15° shift between corresponding secondary windings.

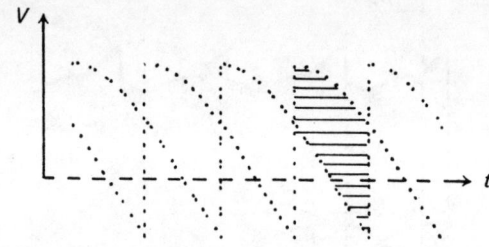

Fig. 75   Larger voltage–time area (shaded) to be supported by an interphase reactor when the rectifiers supplying it operate with phase delay

Whenever more than one convertor is operating on a common supply, each can disturb the supply-voltage waveform and hence modify the operation of the other (Freris, 1967; Mellgren, 1965; Stahl, 1969). In principle, the effect is similar to overlap interference.

5.5   Loads: resistive, inductive and back-e.m.f.

5.5.1   Resistive loads

For a resistive load, the load current is proportional to load voltage. Device currents between overlaps are no longer constant, so that voltage drops and rises occur in the leakage reactances as device currents rise and fall. The determination of output voltage waveform is difficult, as it no longer follows the waveshapes of the supplies. The other major difference is that device conduction into regions of reverse output voltage is impossible. Thus a fully controlled rectifier with a resistive load takes on the characteristic of half control. Fig. 77 compares the output voltage waveforms of a single-phase centre-tap ($q = 2$) circuit with resistive and inductive loads for $\alpha = 0°$ and

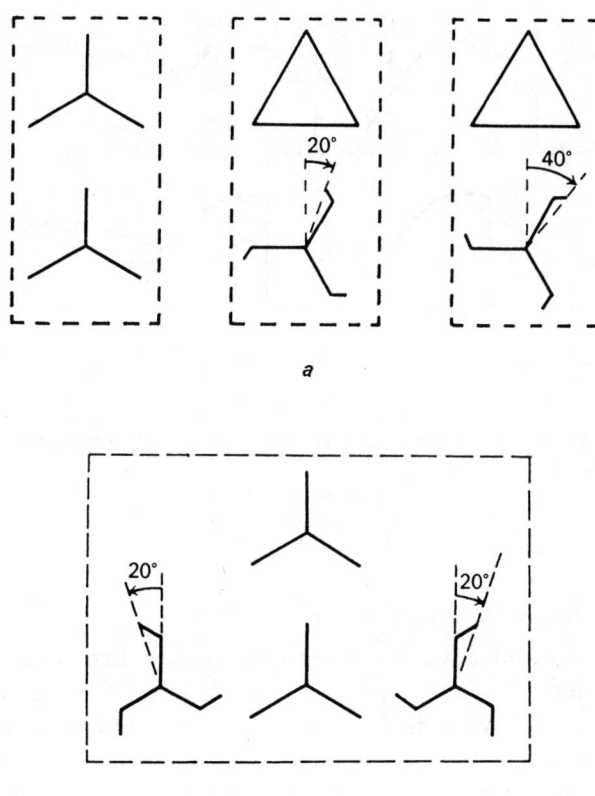

a

b

Fig. 76 18-pulse operation using three 3-phase bridge rectifiers requires three transformers (*a*), or one transformer (*b*), whose secondary voltages are phase-shifted by 20° in both cases

$\alpha = 60°$. For single phase, there is almost a complete lack of overlap, as the current I_d has almost collapsed to zero at the end of the half-cycle. For 3-phase centre tap ($q = 3$), the output voltage, and hence output current, does not fall to zero for $\alpha = 0°$, and thus an overlap exists as for smooth load to current, but of shorter duration, since the instantaneous value of output current is less than its mean value; for $\alpha > 30°$, the output voltage and current do fall to zero, so that overlap ceases to occur.

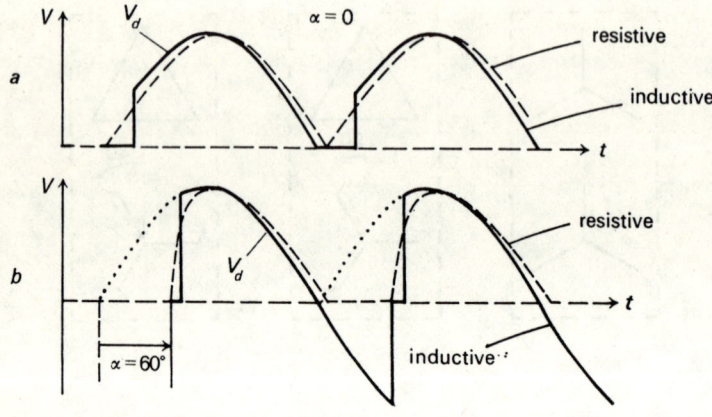

Fig. 77 Comparison of output voltages for single-phase centre-tap rectifier for resistive and inductive loads

a Zero delay angle
b Delay angle $\alpha = 60°$

5.5.2 Inductive loads ($H \neq \infty$)

A practical inductive load, e.g. a field winding, behaves as a series combination of an inductive and resistive load. The factor which governs the behaviour is whether the load current is continuous or not. With continuous current, which usually applies from maximum load current down to a much lower value, the output voltage waveform closely resembles that for smooth current. Transiently, the inductive load is able to maintain continuous current in a fully controlled rectifier in spite of a reversal of the mean output voltage produced by a delay $\alpha > 90°$ (Chapter 6). By this process, an inductive current can be collapsed rapidly by inverting the energy stored in the inductive load and returning it to the a.c. supply.

At low values of output current, the onset of discontinuous current occurs first for the fully controlled rectifier, because reverse voltage–

time area in the voltage waveform collapses the current I_d more rapidly than when a freewheeling action takes place. As the delay angle α of a fully controlled rectifier approaches $\alpha = 90°$, the forward and reverse voltage–time areas approach equality (Fig. 78) for a 3-phase bridge. For larger values of delay, the continuation of conduction into the reverse voltage–time area is, in the steady state, progressively shortened, as less of this area is required to collapse the smaller current I_d to zero. From $\alpha = 90°$ to $\alpha = 180°$, the mean steady-state output voltage V_d is only slightly above zero, by an amount necessary to pass the mean value of the now discontinuous load current I_d through the resistive component of the load.

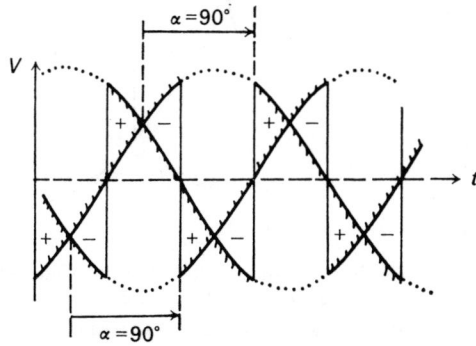

Fig. 78 Equality of positive and negative voltage–time areas at $\alpha = 90°$, illustrated for 3-phase fully controlled bridge

Fig. 79 summarises the steady-state (solid curve) (Borst *et al.* 1966) and transient (dashed curve) control characteristics of all fully controlled rectifiers, centre-tap or bridge, supplying a finite inductive load. Polarity reversal (inversion) is only possible transiently while continuous current persists. The dotted curves are the control characteristics for resistive loads. All half-controlled bridges follow the top dotted curve, which also applies to the single-phase ($q = 2$) centre-tap and fully controlled bridge circuit. The 3-phase centre-tap circuit follows the middle dotted curve, while the 6-phase centre-tap and 3-phase fully controlled bridges follow the bottom dotted curve.

Any fully controlled (or centre-tap) rectifier with a freewheeling diode supplying an inductive load has the same control characteristic as it would have supplying a resistive load.

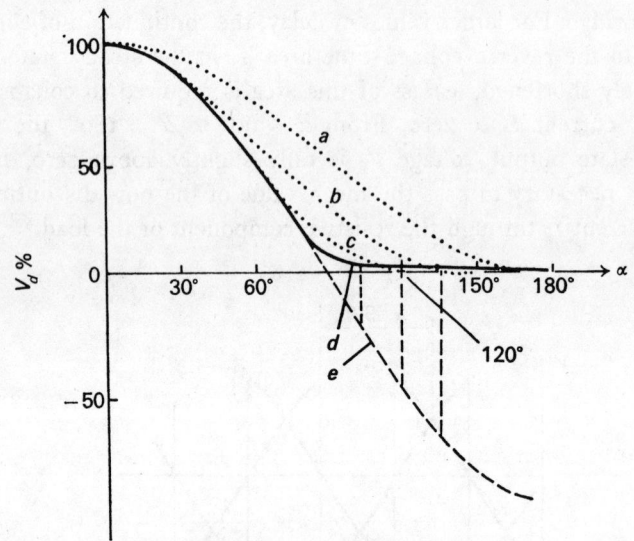

Fig. 79   Control characteristics of all fully controlled rectifiers
for resistive and inductive loads

All half-controlled rectifiers follow top dotted characteristic

c.t. = centre tap; b. = bridge; h.c. = half-controlled

Resistive loads $\begin{cases} (a) \ q = 2; \text{ c.t. and b.; all h.c.} \\ (b) \ q = 3; \text{ c.t.} \\ (c) \ q = 6; \text{ c.t.}/q = 3; \ b. \end{cases}$

Inductive loads $\begin{cases} (d) \text{ discontinuous current, steady state} \\ (e) \text{ continuous current, transient} \end{cases}$

5.5.3   Back-e.m.f. loads

This class includes capacitive and battery loads, and most important, motor-armature loads, embracing the entire field of rectifier variable-speed drives (Vedder and Puchlowski, 1943; Puchlowski, 1945).

The presence of a large capacitor directly in parallel with the output of a single-phase rectifier, though favoured as smoothing for small power supplies for electronic equipment, is not recommended for high-power rectifiers. The capacitor, remaining charged approximately to the peak value of the alternating voltage, demands short-duration high-amplitude pulses of current from the rectifier, giving a high r.m.s. value to the device and supply currents, with consequent derating.

Power rectifiers supplying capacitive or back-e.m.f. loads do so through some smoothing inductance which, at least for full-load conditions, maintains continuous current I_d from the rectifier. For full-load conditions, therefore, the rectifier operation approximates closely to that of the smooth-current analysis presented in Chapters 4 and 5.

The onset of discontinuous current for a fully controlled rectifier with an inductive load occurs for a firing delay α approaching 90°, where the negative voltage–time area has almost reached equality with the positive voltage–time area. For a back-e.m.f. load, however, reverse voltage-time area occurs whenever the instantaneous output voltage of the rectifier is less than the back e.m.f. Thus the likelihood of discontinuous current is much greater, and it can occur at all firing-angle delays. As long as continuous current is flowing, the rectifier output-voltage waveform follows the theoretical waveshape closely, being modified only in so far as variations in I_d produce e.m.f.s in the leakage inductances which add to, or subtract from, the sinusoidal phase voltages. The onset of discontinuous current results in periods of zero current during which no rectifier devices are conducting; the output voltage for these periods of zero current is equal to the back e.m.f. of the load. Thus the load voltage waveform follows that of the rectifier during rectifier conduction, but that of the load back e.m.f. at other times. Fig. 80 shows a succession of halfcycles of output voltage from a single-phase ($q = 2$) rectifier, with successively increasing values of back e.m.f. shown as horizontal lines. A back e.m.f. less than V_d (mean) for the rectifier must produce continuous current (neglecting load resistance). However, substantial pulses of current also flow for back e.m.f.s greater than V_d (mean) where the load inductance is small. These pulses of current are shown in Fig. 80, with the back-e.m.f. line as their zero for clarity. The voltage–time areas which build up and collapse the load current are shaded vertically and horizontally.

A motor armature supplied from a single-phase ($q = 2$) rectifier, having waveforms as shown in Fig. 80, can, at light-load conditions requiring a low armature current, accelerate until its back e.m.f. is close to the peak value of the sinusoidal supply. Applying load to the motor requires additional armature current which can only be obtained if the back e.m.f. (and speed) falls appreciably, in the limit to the mean output voltage of the rectifier. Inherent speed regulation is thus very poor, until continuous current is established.

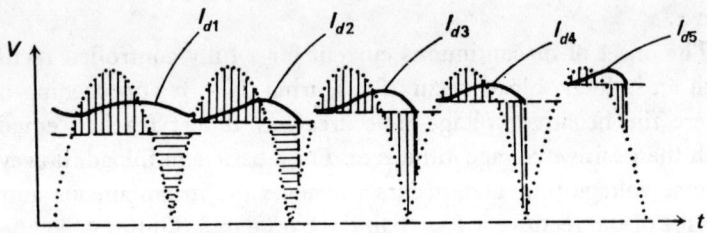

Fig. 80 Output voltage (dotted) of single-phase rectifier feeding a back-e.m.f. load (five values of back e.m.f. shown dashed) via a smoothing inductor

Voltage–time areas across inductor are shaded vertically and horizontally for rising and falling load current I_d. Discontinuous current occurs for all but the lowest back e.m.f.

For a controlled rectifier, further complications arise concerning the correct firing of the thyristors. With continuous current, it is possible to fire the thyristors of the $q = 2$ rectifier at zero delay, with normal overlap behaviour. However, for the discontinuous-current conditions shown on the right-hand side of Fig. 80, an attempt to fire the incoming thyristor at $\alpha = 0°$ is confronted with a reverse-biased thyristor, since the load voltage exceeds the phase voltage. Pulse firing is thus wholly unsatisfactory: the motor would accelerate from rest, drawing continuous current, to high speed; discontinuous current would then set in, making the firing at $\alpha = 0°$ ineffective, so that the motor would coast to rest; correct firing at $\alpha = 0°$ would then again become possible, repeating the cycle of events. Prolonged gate pulses and closed-loop speed (or voltage) control is required for the successful operation of the single-phase rectifier into a motor-armature load.

Polyphase rectifiers, while less susceptible to the above problems, still exhibit an output voltage ripple, particularly at the lower output voltages produced by firing delay. Discontinuous current can therefore occur just as for the single-phase case. The problem of firing the thyristors is less severe; this is clear from Fig. 81, in which a repetition period of output voltage is compared with its mean value for single-phase (Fig. 81 *a*), 3-phase centre tap (Fig. 81 *b*) and 3-phase

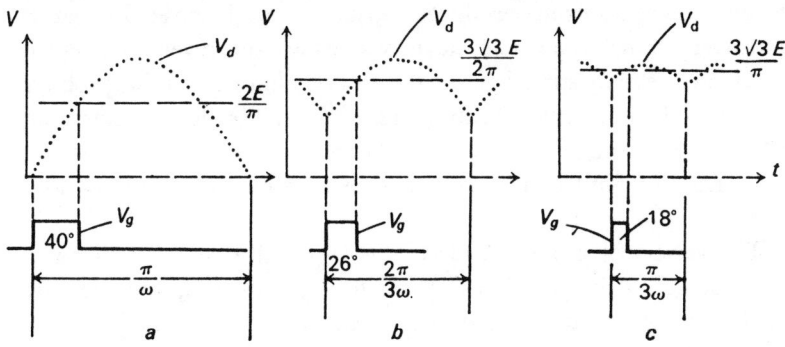

Fig. 81 Showing minimum angular duration of gate pulses V_g, for (*a*) single phase, (*b*) 3-phase, centre-tap, and (*c*) 3-phase bridge rectifiers for correct firing into back-e.m.f. load whose voltage is equal to the mean output voltage of rectifier for continuous current

bridge (Fig. 81 *c*). For a back e.m.f. equal to the mean rectifier output at zero firing delay, the minimum length of the firing pulses must be as shown. Where the back e.m.f. may rise to the peak rectifier-output voltage, the limits for pulse lengths are 90°, 60° and 30°, respectively.

5.6 Rectifier protection

The small thermal mass of a semiconductor junction results in a low thermal time constant. Even with a massive cooling fin, the thermal time constant of a power semiconductor is only of the order of a few minutes. This is in marked contrast to the thermal time constants of other electrical plant, particularly of rotating machines which have a time constant an order of magnitude greater. This is the basic fact which colours the whole subject of overcurrent protection, whether fault protection by fuselinks, mentioned in Chapter 2, or overload protection by circuit breaker or, in the case of controlled rectifiers, by electronic current control. What is a short-duration overload for the motor of a rectifier drive may have to be considered as the continuous rating of the rectifier itself.

Overvoltages on semiconductor devices raise the leakage current often in a very localised area of the junction, so that a relatively small reverse leakage current produces substantial local power dissipation and overheating. Avalanche devices in which the reverse current is much more evenly spread across the junction have eased this problem considerably, but have not entirely removed it. The devices must thus be protected from overvoltage, whether generated within the rectifier by normal or fault conditions, or appearing on the supply or load systems.

The variety of protective circuits is large; all that can reasonably be presented is the principles on which they are based, together with a few examples of those used more frequently.

5.6.1 Overcurrent protection

Diode rectifiers do not have the facility of control, and hence their protection against overcurrent must be performed by external means.

Fuselinks can be used for the two quite separate purposes described in Section 2.4.3. A device fuselink in equipment with a few devices may be used to protect the device against excessive currents or faults, but, in large equipment, it is used solely to isolate a faulty device and thus allow continued operation. The overcurrent protection of such large rectifiers is thus left to a circuit breaker or fuselink in series with the load or supply. The rectifier would consequently be rated to carry the fault current during the clearing time of the breaker, or external fuselinks, without blowing any rectifier fuselinks. Additional reactance is sometimes useful in limiting the fault current to an acceptable value during the clearing time.

Correct rectifier protection (Corbyn and Potter, 1960; Gentry, 1958; Golden, 1968) requires that, whatever the magnitude or duration of the overcurrent, the devices will not be subject to currents exceeding their survival limit. A nonrepetitive current/time survival characteristic appears in the device data sheets, a typical example being shown in Fig. 82; the current/time characteristics of the protective devices, e.g. circuit breaker, fuselinks, reactors etc., must combine to give a characteristic which falls below that of the device, as shown.

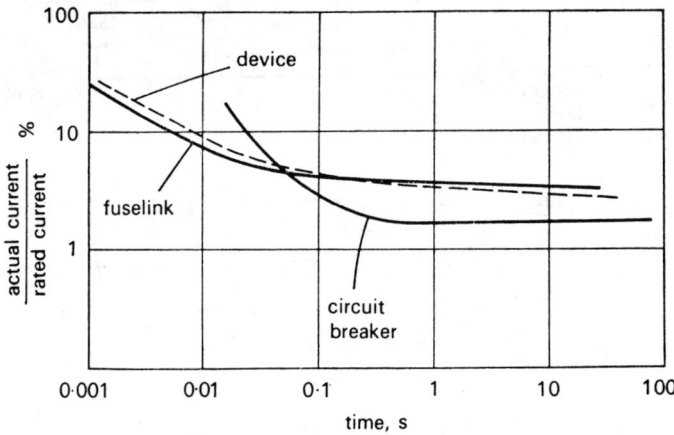

Fig. 82 Survival characteristic for semiconductor device (dashed) which is adequately protected by a fuselink for short-duration high-current faults, and by a circuit breaker for more prolonged overloads

Controlled rectifiers are protected by electronic current-control circuits which sense the device current usually by measuring either output or supply current, and automatically delaying or removing the thyristor firing pulses in the event of an overcurrent. These electronic circuits limit the duration of the fault to a halfcycle or less, for which the supply reactance is usually sufficient to limit the magnitude of the fault current.

Electronic overcurrent protection can take several forms. An electronic trip circuit can be used to remove the firing pulses, which are then reapplied manually or automatically after a short period.

An electronic current-limit circuit can be used, which is an overriding signal to the phase-control circuits, delaying the firing pulses progressively as the current exceeds the preset value. Similar action is obtained by a continuously acting current-control loop for which the

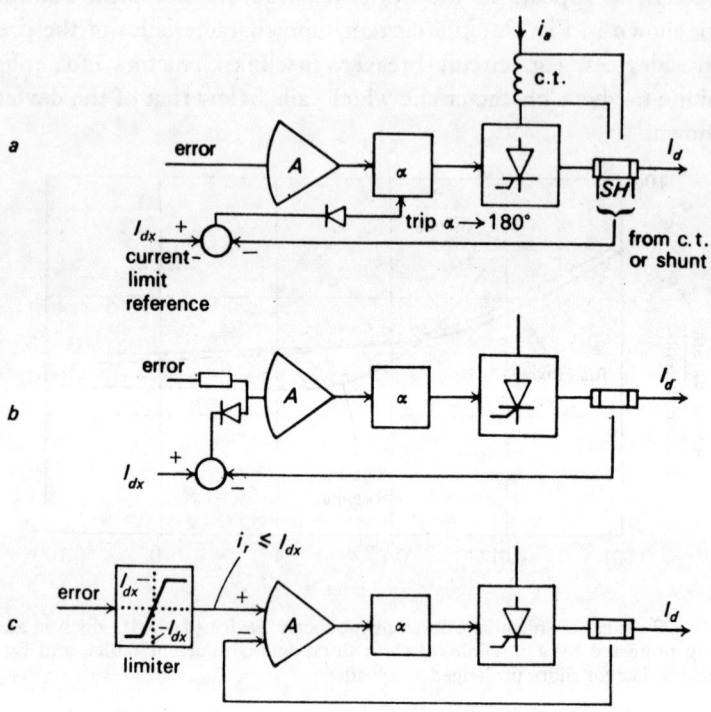

Fig. 83   Electronic current limitation

a Overcurrent trip which fully delays or removes firing pulses from thyristors
b Current-limit circuit which provides corrective signal to override the error signal if output current I_d exceeds limit level I_{dx}
c Current-control loop in which the demanded current i_r is limited to I_{dx}

demanded current signal is limited to a safe value. The continuous operation of the current-control loop makes for easier stabilising, and is now almost universally used for rectifier drives. These three techniques are represented in Fig. 83.

5.6.2 Overvoltage protection

Overvoltages are caused by a sudden release of energy either inside the rectifier equipment or outside it (Corbyn and Potter, 1960; Corbyn, 1965; McColl and Whitaker, 1965). Overvoltage protection is based on the absorption of this energy in a capacitor, or its dissipation in a nonlinear resistive device at a circuit voltage which, although above the normal crest working voltage, does not overstress the rectifier devices. Protection thus demands a significant margin between the device voltage corresponding to the crest of the supply voltage and transient-voltage capability of the device. Factors of safety of 1·5–3 are common.

Supply transient voltages, caused by circuit switching, lightning and fault clearance, are always present, the smaller overvoltages having normally longer durations. The filtering of the longer-duration transients is difficult, and this sets the minimum level for the voltage safety factor at about 1·5. Very much higher transient supply voltages are also present, but their much shorter durations make filtering straightforward.

The energy associated with supply transient is unknown. Information has been gathered over long periods of time for typical industrial situations on which the present factors of safety are based (Bull, 1968).

Internally generated transients have a calculable energy content, and protection circuits can thus be designed with some confidence to cater for these known conditions.

The sudden release of energy is associated with a switching operation, either mechanical or solid-state. Possible sources of released energy are listed below in order of importance:

(a) sudden interruption of transformer, or similar magnetising current, releasing the stored energy in the field
(b) sudden interruption of load current, releasing stored energy in transformer leakage or similar inductances
(c) sudden energisation of the primary of a step-down transformer, with interwinding capacitance, producing a transient secondary overvoltage

(d) motor (back-e.m.f.) load, regenerating an excessive voltage owing to overspeed or rise of field current

(e) sudden energisation of primary, causing secondary winding to 'ring' (self capacitance and inductance resonate), giving a transient secondary voltage approaching twice the normal peak value

(f) sudden interruption of carrier storage current in a semiconductor device releasing energy stored in series inductance as an over-voltage across the device itself; this is limited by a parallel capacitor.

When the protective circuit is a capacitor, it can be placed, with some series resistance to damp ringing, across the transformer secondary, or, in the case of a diode rectifier, across its direct-voltage output. The latter position minimises the power frequency current carried by the capacitor, and allows an electrolytic capacitor to be used where acceptable from other viewpoints: also, in polyphase circuits, only one capacitor is required. In either position, the peak capacitor voltage occurs at the peak of the alternating supply voltage, and this is the worst instant for an interruption of load current, since the capacitor is already charged when called on to absorb the energy stored in the leakage inductances.

For a single-phase rectifier, the worst instant for the interruption of magnetising current is at the end of the voltage halfcycle, when the magnetising current is at its peak value. At this instant, the capacitor may or may not be charged, depending on whether it is connected to the d.c. or a.c. side of the rectifier, and on the nature of the d.c. load. The interruption of magnetising current for a 3-phase rectifier always finds the capacitor(s) precharged to some extent. The capacitance C is found from the stored energy from the relationship

$$W = \tfrac{1}{2}C(V_{C2}^2 - V_{C1}^2) \tag{5.6}$$

where V_{C2} is the voltage across the capacitor corresponding to the peak transient device voltage and V_{C1} is the highest initial voltage on the capacitor.

Controlled rectifiers are economically protected by a small diode bridge and capacitor resistor circuit (Fig. 84a). Where a bridge rectifier is half-controlled, only half as many extra diodes are required to make up the protective bridge. This technique is preferable to placing series RC networks across the rectifier a.c. terminals (Fig. 84 b) since, when so placed, they carry appreciable power-frequency currents, and also modify considerably the commutation behaviour. The time constant of the RC networks is generally short enough to allow discharge and recharge during overlap, giving an additional power dissipation in the resistors, which rises typically to 1–2 % of the rectifier rating for firing-angle delays of 90°.

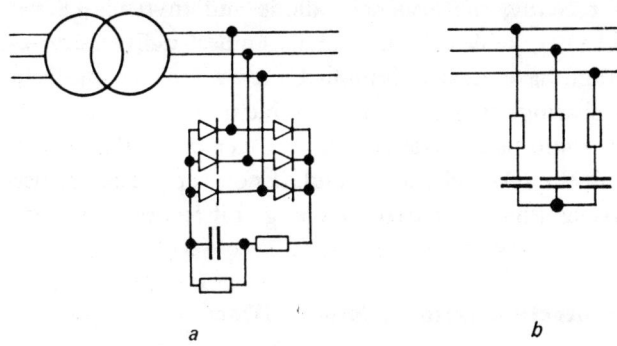

a   b

Fig. 84 (a) Voltage-transient suppressor in which an uncontrolled bridge recti-fier enables a single capacitor, in conjunction with transformer leakage reactance, to attenuate transients in all phases
(b) Transient suppressor in which components carry power-frequency currents

The capacitor requires series resistance to damp the ringing at switch on, which otherwise would allow the capacitor and rectifier voltage to approach twice the peak of the supply voltage. Voltage safety factors of less than 2 make the series resistance essential. The optimum value for the series resistance R is given by

$$R = 2H/C \qquad (5.7)$$

where H is the total series inductance, equal to the leakage inductance for centre-tap rectifiers and twice the leakage inductance for bridge rectifiers.

Nonlinear resistors of thyrite, or more recently selenium (Martin), are also used for rectifier transient-voltage protection. They can be used on the a.c. or d.c. side of a rectifier, or used across the devices themselves. Because they do not conduct significantly at the peak supply voltage, there is little or no power-frequency current. The selenium devices are in fact selenium rectifiers used in their reverse voltage quadrant. For protection on the a.c. side of a rectifier, they must consequently be used in reverse-series pairs. On the d.c. side of a rectifier or across a diode (not thyristor which supports both polarities of voltage), a single selenium protector suffices.

Most recently, the avalanche diode and thyristor (Yanai, 1965; Gutzwiller), capable of substantial nondestructive reverse-current conduction, have greatly simplified the overvoltage protection (and voltage sharing) in power semiconductor equipment. The energy concept is still used to determine whether the avalanche device can absorb all the released energy, and, if not, capacitors are used as for other devices, but to absorb the energy difference.

5.6.3 Protection against di/dt and dV/dt

Where thyristors are included in a rectifier, it is also necessary to ensure that di/dt (Mapham, 1963) at turnon and dV/dt at turnoff as well as through the 'off' state are within the manufacturer's specification.

While protection against di/dt is provided by inductance almost always present on the a.c. side of a rectifier, capacitors for overvoltage protection frequently are a source of high di/dt, when they are connected either to the a.c. side of a rectifier or across devices. When looking for possible sources of high di/dt, it must be remembered that other semiconductor devices are capable of transient reverse carrier-storage conduction following forward conduction, and this behaviour is often responsible for the occurrence of a high di/dt loop. See also the comments in Chapter 2 on series operation of devices.

Protection against dV/dt is achieved with capacitors of 0·1–1 μF connected in parallel with the devices, with a series resistor or inductor to limit di/dt at turnon. In rectifier circuits, dV/dt occurs at the end of overlaps when the voltage rises suddenly from an overlap voltage to a phase voltage; where a reverse-parallel rectifier is connected, it can consequently experience a high dV/dt. The thyristors of a synchronous invertor can similarly experience high dV/dt (Rice and Nickels, 1968).

6 NATURALLY COMMUTATED INVERTORS AND REGULATORS

6.1 Introduction

Natural commutation occurs in all convertor and regulator circuits in which the supply voltage either produces thyristor turnoff at a naturally occurring current zero, or enables the firing of one thyristor to turnoff the preceding one. Natural commutation contrasts with capacitor commutation, in which a precharged capacitor performs the above function of the supply voltage.

Rectifiers commutate naturally when conduction is transferred from one device to the next by the supply voltage as the incoming device begins conduction. This Chapter groups together other convertors and regulators which rely on natural commutation.

The synchronous invertor is nothing more than a fully controlled rectifier (convertor) operated with a delay angle $\alpha > 90°$, for which the mean output voltage of the convertor has a reversed polarity. As the current cannot change direction, the power flow is also reversed and the synchronous invertor converts the d.c.-system power into a.c.-system power at the voltage and frequency of the existing a.c. system. Synchronous inversion is used in h.v. d.c. systems (Cory, 1965; Adamson and Hingorani, 1960; *IEE Conf. Publ.* 53, 1969), and in rectifier drives when substantial or frequent overhauling loads make regeneration an economical alternative to dynamic braking.

The cycloconvertor converts alternating current of one frequency into alternating current at another, usually lower, frequency. A fully controlled rectifier bridge with a suitably modulated delay angle produces a sinusoidally modulated d.c. output. Two reverse-parallel rectifiers are required to form both halfcycles of current, and three rectifier pairs are required for a 3-phase output. Loads of any power factor can be supplied, as the reverse-parallel rectifier connection allows either direction of load current to flow for either polarity of the load voltage. Reverse power flow is also possible.

Historically, cycloconvertors were used for the generation of $16\frac{2}{3}$ Hz from 50 Hz for railway-traction supplies (Rissik, 1935). The fixed frequency ratio allowed a greatly improved output waveform to be obtained. Now, the applications of interest require a variable frequency, usually for variable-speed operation of a.c. motors.

A.C. regulators produce a variable-voltage (and variable-waveform) alternating output from the fixed-voltage sinusoidal supply. Two reverse-parallel thyristors (or a triac) are normally required for the formation of both halfcycles of the output voltage waveform, although, in polyphase applications, a thyristor–diode pair can also be used. The criterion is the required absence of a direct-current component in the supply.

The only notable application for this technique with mercury-arc tubes (ignitrons) has been in welding control, where the reverse-parallel devices act merely as switches; thyristors are now similarly used for welding control.

The main growth area is for a.c. regulators which provide smooth voltage control. Applications range from lamp dimming (Storr Best, 1965), to electric-furnace control (Pollard *et al.* 1969), (20 W–20 MW). The use of thyristors and triacs as static contactors (Feltbower, 1969) is also increasing as the devices become cheaper, in applications where frequent operation of an electromechanical contactor would involve unreliability or frequent servicing.

6.2 Synchronous invertors

Inversion, or the conversion of d.c. power into a.c. power, has already been introduced when discussing centre-tap and fully controlled bridge rectifiers operating with a firing delay $\alpha > 90°$ which, in combination with continuous direct current, produces a reversed output voltage, and hence a reversal of power flow.

Half-controlled bridge rectifiers and rectifiers with freewheeling diodes are incapable of inversion, as the reverse voltage necessary for inversion cannot be developed across the bridge or freewheeling diode.

Whereas rectification is inherently a 'safe' process, as shown by the ability of the rectifier to continue operating for all currents up to the short-circuit value, the inversion process fails suddenly (Arrillaga and Galanos, 1962) when the sum of the delay angle and the overlap angle approach 180°, or more precisely when

$$\alpha + \mu \geqslant 180° - 360 f t_{off} \qquad (6.1)$$

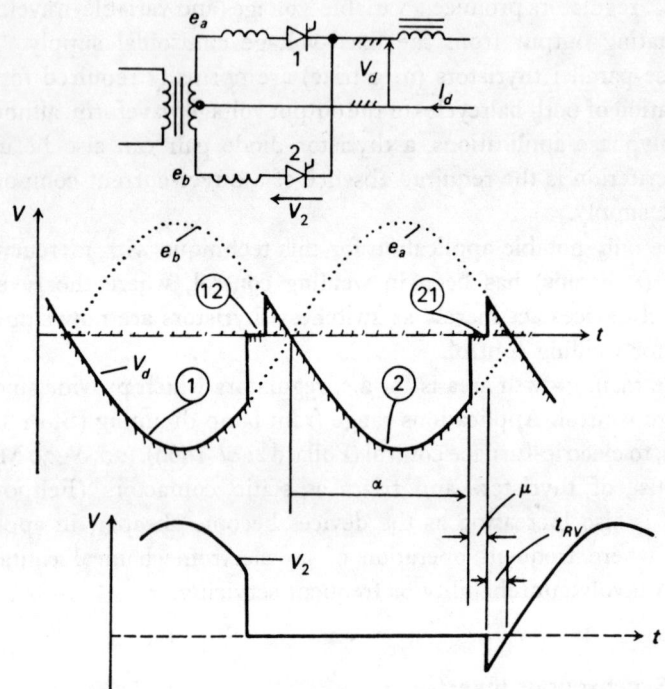

Fig. 85 Circuit and voltage waveforms for single-phase centre-tap invertor (q = 2)

Lower waveform illustrates short reverse-voltage time t_{RV} during which thyristor 2 must recover its forward blocking capability

Fig. 85 shows the operation of a single-phase centre-tap invertor, with the associated voltage waveforms. While V_d now represents the mean input voltage, it is emphasised that, as for rectifiers, the analysis and practical operation are based on constant smooth current. The input-voltage waveform is therefore wholly determined by the invertor alternating voltage, leakage inductance and firing angle, as is the rectifier output voltage waveform. When an invertor is to be supplied from a constant-voltage d.c. supply, it is essential to interpose a d.c. smoothing inductor to enforce smooth current by supporting the momentary differences in potential between the supply and the invertor. Correct operation is thus established with the waveforms as shown.

If the applied direct voltage exceeds the back e.m.f. of the invertor, smooth continuous direct current flows, giving the invertor the waveforms illustrated (or readily drawn for other circuits based on rectifier operation, but with the delay angle $\alpha > 90°$). Smaller applied direct voltages are insufficient to establish smooth current, and a discontinuous current flows, rising when the instantaneous invertor voltage is less than, and falling when it exceeds, the applied direct voltage.

For continuous smooth direct current, the voltage waveforms, and particularly the waveform across a thyristor, show why commutation failure occurs. The period t_{RV} of reverse voltage across the thyristor after conduction ceases is

$$t_{RV} = (\beta - \mu)\, T/360 \qquad (6.2)$$

using the angle of advance β ($= 180 - \alpha$) in preference to the delay angle. If t_{RV} becomes less than t_{off}, owing to an insufficient angle of advance or to an excessive overlap angle, correct turnoff will not occur, and commutation failure results. Thus eqn. 6.2 becomes

$$\beta - \mu < 360 f t_{off} \qquad (6.3)$$

As an example, the behaviour of a 3-phase bridge after a commutation failure is examined, assuming that the firing sequence and angle of advance are unaltered. Fig. 86 shows the commutation failure when, after firing 6, 4 fails to turn off before the a–b phase crossover. After the crossover, the current in 4 rises again at the expense of the current in 6 until 6 turns off, leaving the anode output terminal following the potential of phase a while 4 conducts alone. Meanwhile 5 has been conducting alone in the cathode halfbridge. The failure to commutate gives an instantaneous invertor back e.m.f. which is less than would have prevailed after a successful commutation; so the direct current I_d must rise, which inevitably brings another commutation failure between 5 and 1 (assuming no change in β). Although, when 1 is fired, 1 and 4 form a short-circuit path, the direct-current rise is still limited by the d.c. inductor. Before the phase crossover $c-a$, the current in 5 decays and that in 1 rises: but, after the crossover, the reverse takes place, ultimately leaving 5 and 4 as the only conducting devices. The cathode output terminal of the invertor has

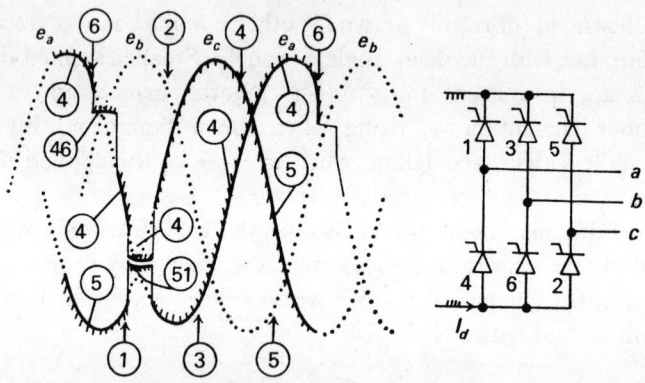

Fig. 86 Events following the commutation failure 4 to 6 of a 3-phase bridge-invertor, showing the collapse of invertor back e.m.f., a halfcycle in which the bridge polarity assists current flow and a third of a cycle in which the invertor back e.m.f. increases, terminated by another commutation failure. Ringed numbers with arrows outside the waveforms indicate firing instants

now become positive with respect to the anode output terminal. After another halfcycle, just before the next $c-a$ crossover, the firing pulse for 4 occurs when 4 is already conducting, so that no overlap takes place. After this crossover, the bridge voltage again becomes negative; the next attempt at commutation arises when 6 is fired, which will again fail, repeating the above sequence. It is clear that the bridge voltage of zero mean value is an alternating voltage equal to the line–line supply voltage.

Minimisation of the reactive-power demand of an invertor (Gardner and Fairmaner, 1968) is a good reason for operating with the smallest possible angle of advance, but this demands fast control of β if commutation failures due to transient rises in direct current or drops in alternating voltages are to be avoided (Reeve, 1967; Ito and Sekine, 1968).

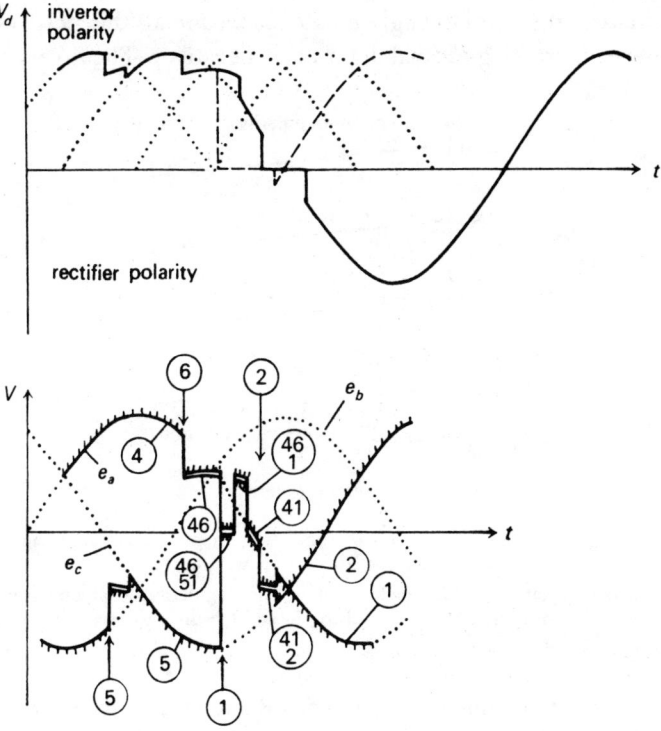

Fig. 87 Direct-voltage waveform of 3-phase bridge invertor following a commutation failure when there is no change of firing pattern (as for Fig. 86) (shown solid), and when the firing pattern is altered (shown dashed) in accordance with the voltage waveforms

If every commutation is monitored and a failure rapidly detected, it is possible to alter the firing sequence (Reeve and Burdett, 1969) in a way which avoids the full halfcycle of the positive bridge voltage. The voltage waveforms are given in Fig. 87 without explanation. The rise of direct current is reduced, and normal operation can be resumed a halfcycle sooner.

127

For h.v. d.c. systems, a bypass valve is used which performs a function similar to the early firing of thyristor 1 in Fig. 87, allowing the bridge itself to recover its blocking capabilities.

The single-phase V_d waveform of Fig. 85 emphasises that V_d is negative, but it has a short positive area after each commutation. As I_d increases, the overlap angle μ extends, reducing this positive area and hence increasing V_d (making it more negative). The invertor thus

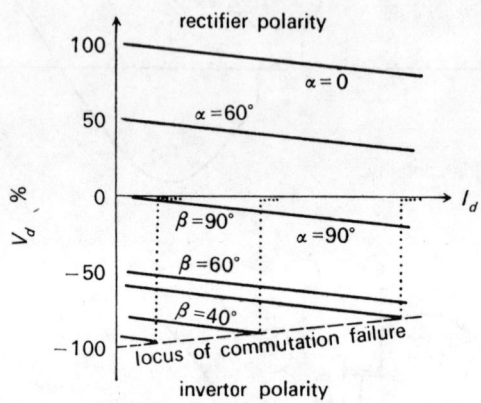

Fig. 88   Rectifier and invertor regulation characteristics showing the rise of invertor back e.m.f. with current I_d

Invertor commutation failure occurs when the regulation characteristic meets the locus of commutation failure, resulting in sudden drop to zero (shown dotted) of the invertor back e.m.f.

has a rising regulation characteristic. Rectifier and invertor regulation characteristics are expressed together in Fig. 88, which shows the natural transfer from rectification to inversion when the delay angle α exceeds 90°, and a typical commutation limit. The slope of the commutation-limit line is the same as that of the regulation characteristic, as the voltage–time area lost for a given increase of the current I_d (when $\beta - \mu = 360 f t_{off}$ for the invertor) is the same in both cases.

The analysis of invertor circuits is identical to that of centre-tap and fully controlled bridge rectifiers, except that V_d and I_d become the input voltage and current, respectively, and the delay $\alpha > 90°$.

Thus the following equations for rectification under these conditions yield an input voltage V_d of reversed polarity which increases with the current I_d:

Single-phase centre tap $(q = 2)$:

$$V_d = \frac{2E}{\pi} \cos\alpha - \frac{I_d\omega h}{\pi} \qquad (6.4)$$

or
$$V_d = -\frac{2E}{\pi} \cos\beta - \frac{I_d\omega h}{\pi} \qquad (6.5)$$

3-phase centre tap $(q = 3)$:
$$V_d = -\frac{3\sqrt{3}E}{2\pi} \cos\beta - \frac{3I_d\omega h}{2\pi} \qquad (6.6)$$

Single-phase bridge $(q = 2)$:
$$V_d = -\frac{4E}{\pi} \cos\beta - \frac{4I_d\omega h}{\pi} \qquad (6.7)$$

3-phase bridge: $\quad V_d = -\frac{3\sqrt{3}E}{\pi} \cos\beta - \frac{3I_d\omega h}{\pi} \qquad (6.8)$

Eqns. 6.4–6.8 all have the overlap limitation of eqn. 6.3.

The current waveforms injected into the a.c. system by an invertor have the same general shape as those drawn by a rectifier operating at $\alpha = \beta$. However, whereas the fundamental rectifier current waveform lags behind the voltage waveform by an angle ϕ where $\cos\phi$ is the displacement factor, the fundamental component of the current waveform of an invertor leads the voltage by the same angle ϕ. The injection of a leading current into an a.c. system requires the same measures to correct the power factor as the drawing of a lagging current from it. Thus an invertor consumes lagging reactive power while supplying power to an a.c. system.

The detailed current waveforms of a rectifier and invertor differ slightly as shown in Figs. 89 a and b, which show the currents for a 3-phase bridge. For a rectifier, di/dt increases as overlap proceeds, since the supply voltage causing commutation is rising. For an invertor, the voltage causing commutation is decreasing as the phase crossover at which $\beta = 0$ is approached.

The equivalent circuit of an invertor at its d.c. terminals includes a back e.m.f. and a series resistance, but the voltage drop across the series resistance represents lost voltage–time area across the leakage inductance, and thus does not involve any power loss. For continuous direct current, the back e.m.f. is given by the first terms in eqns. 6.4–6.8, and the voltage drop across the resistance by the second terms. A diode is also present (Fig. 90) to represent the unidirectional current-flow characteristic of the invertor. The maintenance of smooth current from the direct-voltage source in spite of the voltage fluctuation of the invertor requires an inductance H to give

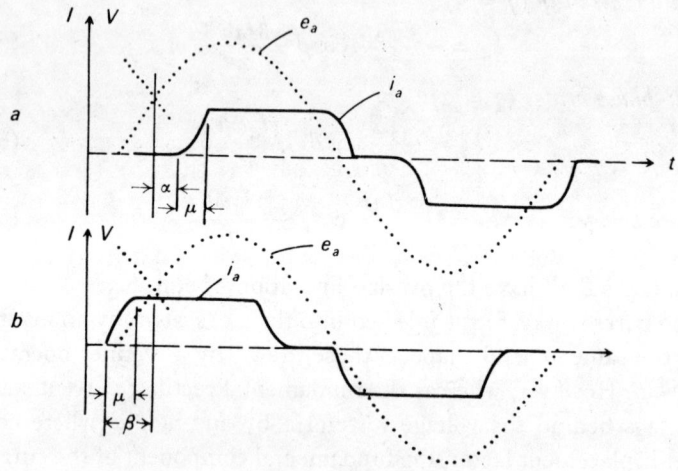

Fig. 89 Comparison of alternating-current waveforms
for 3-phase bridge (*a*) rectifier and (*b*) invertor

a d.c.-circuit time constant which is long compared with the invertor repetition period.

Having set up the equivalent circuit of Fig. 90, it is instructive to examine the behaviour of the invertor for a progressively rising source voltage. The behaviour will be governed largely by the way in which firing signals are applied to invertor thyristors. Assume first that prolonged pulses are used, lasting for the whole of the normal conduction period. In a bridge invertor, two thyristors are then always receiving gate signals, one in each row, thus establishing a circuit through the bridge. This assumption allows direct current to build

130

up as soon as the source voltage exceeds the lowest instantaneous back e.m.f. of the invertor.

For a single-phase centre-tap invertor (operating at $\beta = 20°$, say), the instantaneous back e.m.f. falls below zero twice per cycle, and hence, even for zero source voltage, there will be a momentary rise and fall of direct current (Fig. 91) twice per cycle (as for a rectifier operating at $\alpha = 160°$ into an inductive load).

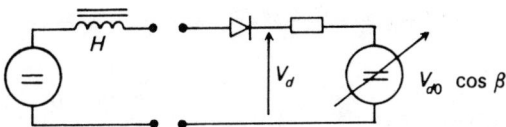

Fig. 90 Equivalent circuit for invertor including diode (unidirectional current property), resistor (simulates voltage drop in leakage inductances) and variable back e.m.f. governed by angle of advance β

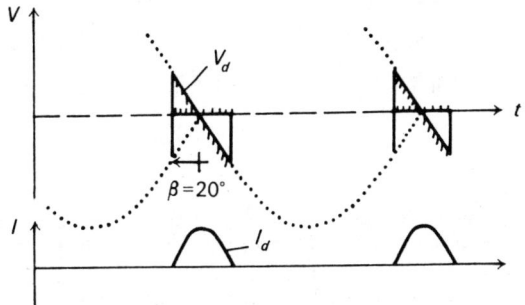

Fig. 91 Short pulses of current flowing in a single-phase centre-tap invertor for zero applied direct voltage

The rise of current occurs immediately after commutation when the instantaneous invertor back e.m.f. is reversed (= same polarity as a rectifier)

A large time constant for the d.c. circuit will keep the pulse amplitude of discontinuous direct-current pulses very small. As the source voltage rises, the voltage–time area building up I_d increases, so that proportionally more area is also required to collapse I_d. When the source voltage reaches the mean invertor back e.m.f. ($2E/\pi \cos 20°$ in this case), the areas for buildup and decay are equal, and a further rise of the source voltage establishes a continuous current I_d. Once continuous current is established, only the d.c.-circuit resistance is present to limit current, and a small rise on the source voltage produces a large increase of current. The characteristic for single-phase invertors is shown dotted in Fig. 92.

3-phase invertors operating at small angles of advance have an instantaneous back e.m.f. which is never less than a minimum value. The source voltage must therefore exceed this minimum value before

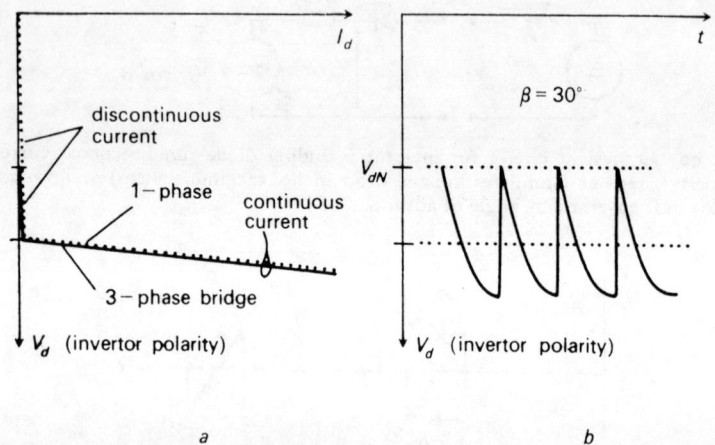

a   b

Fig 92 (*a*) Regulation characteristics, including the low-discontinuous-current behaviour for single-phase invertors (dotted) and for the 3-phase bridge invertor (solid)

(*b*) Invertor back–e.m.f. waveform showing the minimum instantaneous back e.m.f. V_{dN} for $\beta = 30°$

No invertor current flows for applied voltages less than V_{dN}

any current can flow. From this minimum voltage to the invertor, mean back e.m.f., discontinuous current will flow as for the single phase, followed by a sharp rise in continuous current when the source voltage exceeds the invertor back e.m.f. The characteristic for a 3-phase bridge invertor operating at $\beta = 30°$ is shown solid in Fig. 92.

When an invertor is fired by short pulses, discontinuous current can only build up when the source voltage exceeds the instantaneous invertor back e.m.f. during the firing pulse interval. At all other times, no thyristors are receiving pulses, and hence the invertor is an open circuit. As the invertor instantaneous back e.m.f. at the firing instant is at its minimum, the discontinuous-current behaviour is similar for long or short firing pulses. Bridge invertors fired by short pulses must have two thyristors fired at each firing instant to establish the circuit through the invertor. For a 3-phase bridge invertor, the gate pulses (V_g) are illustrated in Fig. 93.

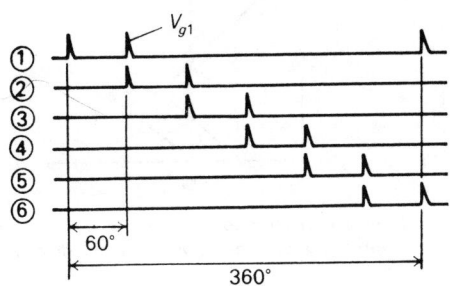

Fig. 93 Gate pulses for 3-phase bridge, showing repeated pulse 60° after the first to establish or re-establish invertor current if it is discontinuous, by firing two thyristors, one in each half bridge simultaneously

The difference in invertor behaviour for short and long pulses is apparent for a rapidly changing source voltage. With long pulses, current starts to flow as soon as the source voltage exceeds the back-e.m.f. waveform of the invertor; with short pulses, current starts to flow at the first firing instant at which the source voltage exceeds the invertor back e.m.f., and this could be a few milliseconds later (Fig. 94).

6.3 Cycloconvertors

6.3.1 Output frequency lower than supply frequency

Cycloconvertors were originally developed as frequency changers for rail-traction supplies at $16\frac{2}{3}$ Hz in the 1920s and 1930s (Rissik, 1935). A fixed frequency ratio of one-third allowed synchronous operation, and, by a suitable choice of 50 Hz voltages, an output waveform close to a sine wave was obtained (Fig. 95). As the present

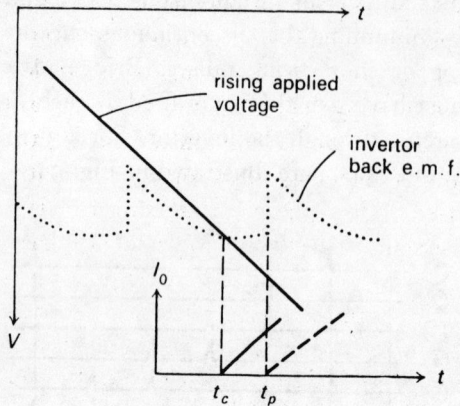

Fig. 94 Showing at t_c an earlier start for invertor current I_d for a rapidly rising applied voltage when continuous gate firing is used, when compared with t_p for pulse firing

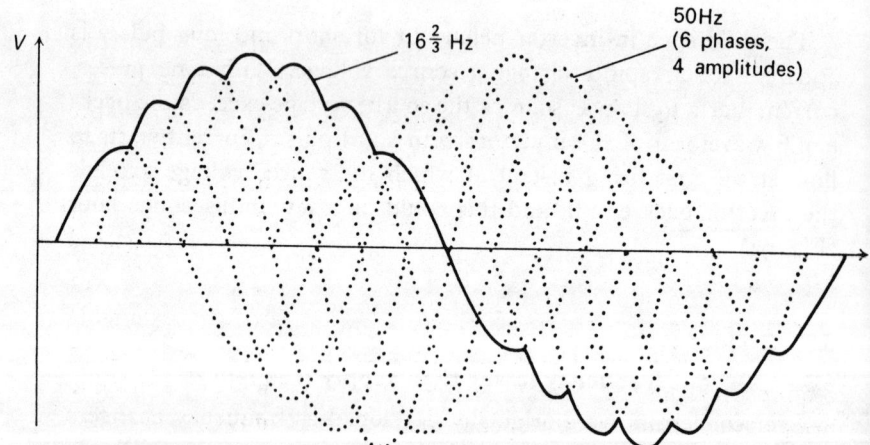

Fig. 95 Output voltage of a 50–16⅔ Hz synchronous cycloconvertor showing the close approximation to a sinusoidal waveshape with correctly chosen 50 Hz voltage amplitudes

growing interest in thyristor cycloconvertors is for variable-frequency and variable-voltage supplies for variable-speed a.c. motors, non-synchronous operation is required which has a poorer voltage waveform (Heck and Meyer, 1963; Lawson, 1968; Bowler, 1965).

To establish a rotating field in an a.c. motor, a variable-frequency supply of at least two phases (Hamilton and Lezan, 1967) is required: a 3-phase supply is obviously most convenient as it allows standard motors to be used. The 3-phase output convertor is somewhat more difficult to understand than that giving a single-phase output, owing to the interactions which can arise between one phase and another. Consequently the 3–1 phase cycloconvertor is described first (Amato, 1966). Two such convertors with 90° phase displacement will power a 2-phase motor, or three with 120° phase shift a 3-phase motor.

The 3–1 phase cycloconvertor is most easily understood for a low output frequency. It comprises two fully controlled rectifiers, centre-tap or bridge, with their outputs connected in reverse parallel. Each rectifier handles one direction of the load current, but is capable of rectifier operation to build up this load current, and of invertor operation to reduce it. Thus, by modulating the delay angle sinusoidally about a mean value of 90°, a sinusoidally modulated output voltage (and hence current) is obtained as shown in Fig. 96 for two 3-phase bridges, using (for simplicity of drawing) triangular modulation. The amplitude of the modulation governs how much the delay angle is reduced towards zero, and hence the peak output voltage. The voltage wave contains an appreciable harmonic content, comprising portions of the mains-supply voltage waveform, but the load inductance ensures that the load current is appreciably smoother.

It is obvious from Fig. 96 that line–line short circuits are possible through one thyristor in each bridge. However, only one bridge is required for duty at a time; so the firing signals for the inactive bridge are blanked off in response to a load-current direction detector. When the load current of the active bridge falls below the detector level, the pulses are blanked from the active bridge (Bland, 1967); and, after a suitable delay to ensure that all thyristors have turned off, the blanking is removed from the other bridge, which then becomes active. The voltage waveforms show clearly how one bridge acts first as a rectifier, and then, for the first part of the reverse halfcycle of voltage, as an invertor.

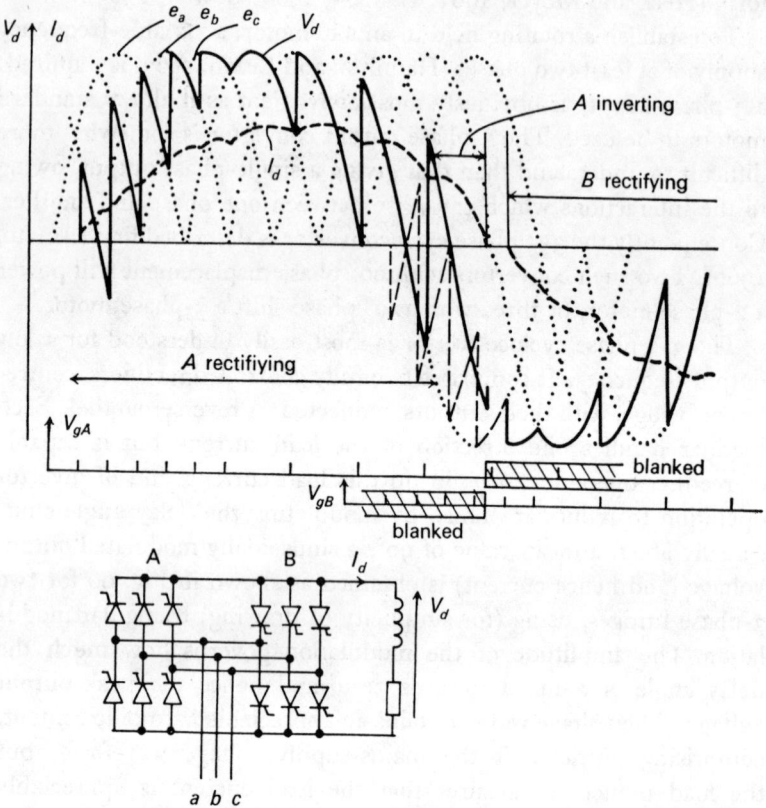

Fig. 96 Nonsynchronous (phase-controlled) cycloconvertor showing output voltage for bridge A rectifying and inverting, during which the gate pulses V_{gB} to bridge B are blanked. When I_d reaches zero, the blanking is reversed, bringing bridge B into rectifying operation and blanking the gate pulses to bridge A

As the output frequency of a cycloconvertor is raised, the number of segments of the supply frequency sine wave per halfcycle of output become fewer, so that the harmonic content of the voltage waveform increases. The upper frequency limit is governed largely by the efficacy of the logic-control circuits and the current-direction indicators. The task of these control circuits is complicated by load-current reversals which, owing to the voltage ripple present, cross the zero-current axis three or more times at the end of each current half-cycle, rather than just once. It is claimed, however, that cyclo-converter operation has been achieved at output frequencies up to and beyond that of the supply frequency (Bowler, 1965). The output waveform at a frequency approaching the supply frequency has

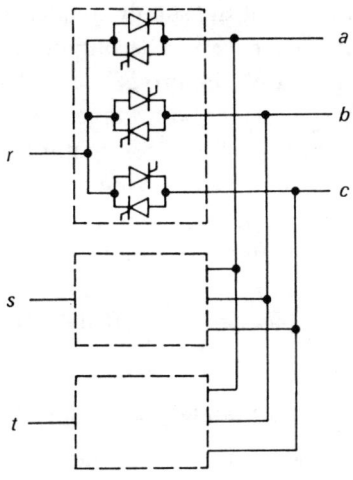

Fig. 97   3-phase–3-phase cycloconvertor using 18 thyristors

a high harmonic content, and a beat-frequency component also becomes apparent. The tolerance of the motor to the worsening waveform limits the practical upper frequency to about 30 Hz for a 50 Hz supply.

The use of three separate cycloconvertors phase-shifted by 120° to supply a 3-phase motor requires 36 thyristors, each cycloconvertor using two reverse-parallel 3-phase bridges. A 3-phase –3-phase cyclo-convertor is possible which uses only 18 thyristors, with the circuit shown in Fig. 97. Here each input line is connected to each output line via a reverse-parallel thyristor pair. If the supply neutral and load neutral are connected, the circuit behaves as three reverse-parallel centre-tap rectifiers, each supplying one phase of the load.

137

When viewed this way, it is apparent that the phase-voltage ripple is at half the frequency obtained when using two reverse-parallel 3-phase bridges, and the neutral link prevents any interaction between the output phases. With the neutral link removed, the other phases contribute a phase-displaced ripple voltage to the phase in question, giving a higher ripple frequency across the phase than with the neutrals linked.

The cycloconvertor is the most flexible of the static frequency convertors. As well as voltage control and a range of frequency control up to typically two-thirds of the supply frequency, it is also possible to reverse the phase sequence of the output without any physical change of power or control circuitry. The cycloconvertor transfers active and reactive power in both directions, allowing a driven machine to operate at any power factor, motoring or generating. It is therefore particularly useful for reversing drives in arduous situations where the robustness of the induction motor makes it a desirable alternative to the more common d.c. motor. The other notable application of cycloconvertors is for obtaining a fixed output frequency and voltage from a variable-speed aircraft alternator (Chirgwin, 1965).

6.3.2 Output frequency higher than supply frequency

'Cycloinvertor' is probably a better description for this circuit, which converts from the supply frequency to a higher output frequency without specifically using a d.c. link. The arrangement is commonly used for induction heating, and was developed with mercury-arc valves. The circuit effectively treats each halfcycle of the alternating supply voltage as a temporary d.c. supply for the invertor. As the next supply-voltage halfcycle becomes more positive, a fresh pair of thyristors take over the inversion duty into the same load components.

The inversion process does not use natural commutation, but uses capacitor commutation described in Chapters 7 and 8. The induction heating coil is resonated with a parallel or series capacitor, and the thyristor is turned off after each resonant half period either by parallel- or series-capacitor turnoff techniques. The capacitor performs the two functions of correcting the power factor of the heating coil and giving the resonant load circuit a leading power factor whereby inverter commutation can occur. As the inductance of the heating coil can vary by 30% or more during a heating cycle, either the capacitance or the frequency must be adjusted to maintain resonance; the latter is easily achieved with thyristor equipment.

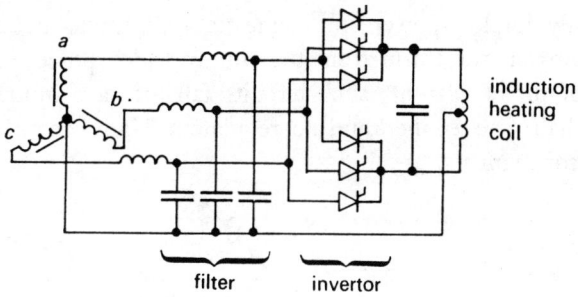

Fig. 98 Cycloinvertor in which the most positive 50 Hz supply phase acts as a temporary d.c. supply for an invertor using capacitor commutation

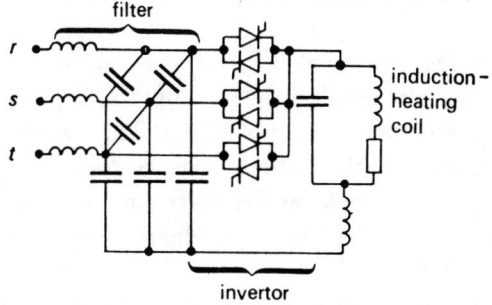

Fig. 99 Alternative to Fig. 98 which avoids the supply transformer and the centre-tapped induction heating coil

Two typical circuits are shown in Figs. 98 and 99. The first is easier to understand. The pair of thyristors, fed from the most positive phase, invert into the resonant load circuit. The second circuit, which does not require a centre-tapped resonant coil or a supply transformer, takes one halfcycle of resonant current from the most positive phase, and the other halfcycle from the most negative phase (Dewan and Havas, 1969).

139

6.4 A.C. regulators

6.4.1 Classification

The variety of a.c. regulator circuits, like that of force-commutated invertors circuits, is so great as to defy classification. Some circuits, however, are excluded for the purposes of this book, being appropriate only for low-power and domestic applications because they draw a current waveform from the supply which possesses a component of direct current; such circuits fall into a nebulous region between halfwave rectifiers and a.c. regulators. Many examples appear in thyristor manuals.

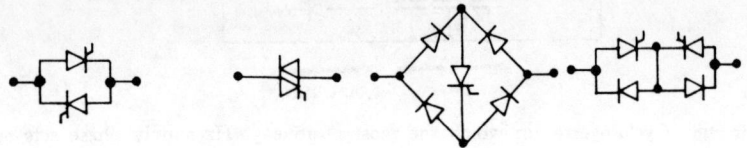

Fig. 100 Four forms of bidirectional a.c. switch
All but the second (triac) use two or more devices

A true a.c. regulator is defined as one drawing an alternating current from an alternating-voltage supply and producing a controllable alternating voltage across a load. The essential component in such a regulator is the bidirectional a.c. switch, which can be obtained using two reverse-parallel thyristors or a triac or a variety of circuits employing thyristors and diodes. Fig. 100 shows some examples.

Any of the switch configurations in Fig. 100, when interposed between a single-phase supply and a resistive load, produce an output-voltage waveform as shown in Fig. 101, depending on whether delay-angle control (Fig. 101 *a*) or integral-halfcycle (i.h.c.) control (Fig. 101 *b*) is used (Knight, 1960; Borst *et al.* 1966; King, 1965; Galloway).

Thus the first distinction has been introduced. Phase control deals identically with every halfcycle (in the steady state) and possesses the same basic shortcomings as in rectification: first, the current wave is delayed with respect to the voltage wave in the process of reducing the output voltage, degrading the power factor; secondly, the nonsinusoidal load voltage at the reduced output gives rise to non-sinusoidal load and supply currents. Similar corrective measures to those required for high-power controlled rectifier circuits are thus also required for delay-angle-controlled a.c. regulators. The harmonic content of the alternating current is not identical to that of a rectifier, since, for a resistive or somewhat inductive load, the load (and supply) current will not be a rectangular flat-topped wave but an asymmetrical rounded waveshape including, in general, all harmonics (Shepherd, 1966).

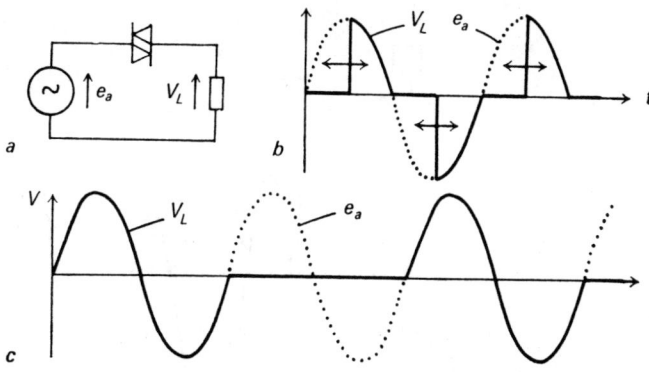

Fig. 101 Single-phase a.c. regulator (a), its output voltage waveforms for a resistive load with (b) delay-angle control and with (c) integral-half-cycle (i.h.c.) control, drawn for one cycle on, one cycle off

I.H.C. control avoids some of the disadvantages of delay-angle control, particularly the high rates of change of current which occur when firing into a resistive load at the crest of the waveform (Lyall, 1969). Harmonics of the supply frequency are minimal, and consequently the interference on adjacent circuits is greatly reduced. Instead, subharmonics are present; e.g. a waveform with every alternate cycle missing has a fundamental component at half the supply frequency. While acceptable behaviour can be obtained for 1 cycle out of 2 (or 2 out of 3, 1 out of 3, 3 out of 5, 2 out of 5 etc.), operation close to the maximum output or close to zero output requires the odd cycle or halfcycle missing or present out of a large number; hence the subharmonic frequency becomes quite low, low enough to produce cyclic voltage drops which appear as a flicker in electric lighting fed from the same supply. The worst flicker frequency is about 8 Hz, at which 1 % voltage changes are detectable in low-power tungsten lamps.

Another classification of a.c. regulators can now be presented. In the simplest 'total' a.c. regulator, the output voltage is either the

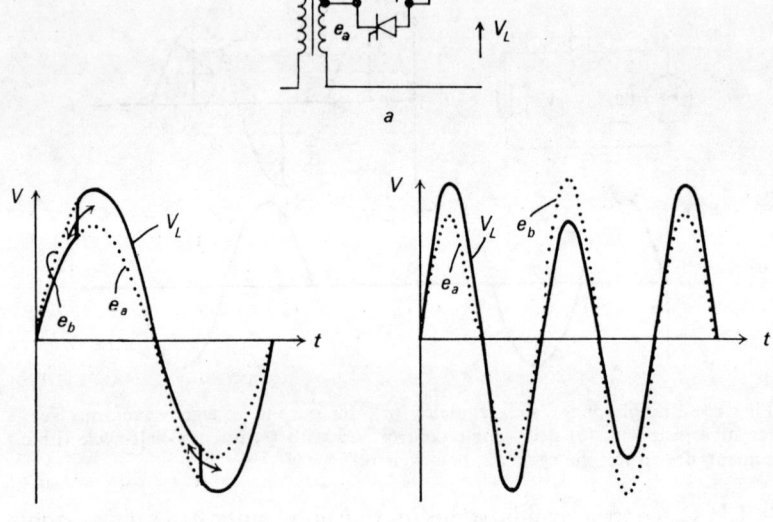

Fig. 102 Differential a.c. regulator (*a*) in which the output voltage V_L for delay-angle control takes the value e_a for the first part of each halfcycle and e_b for the latter part; for i.h.c. control, (*c*) shows the output voltage for $V_{DM} = (e_a + e_b)/2$

input voltage or zero. An important class of differential a.c. regulators exists for which the output voltage is switched between two input voltages rather than between one value and zero (Thompson, 1968; Lloyd, 1969; Borst *et al.* 1966). The same earlier techniques of delay-angle control or i.h.c. control can be applied, as indicated in Figs. 102*b* and *c* showing the output-voltage waveforms for these two control methods applied to the circuit of Fig. 102*a*.

It is immediately apparent that, although the control range of the above regulators is less $(V_{Lb} - V_{La})$ compared with V_{Lb}, their output-voltage waveform, and hence the current waveform, is much less distorted. Transformer-tap changing (Light and McVey, 1967) is an obvious application for differential a.c. regulators, with (Roberts and Ashman, 1969) or without a selector switch to extend the range of control. An effect similar to differential regulation is obtained by series-connecting a low-voltage total a.c. regulator and an uncontrolled voltage, yielding waveforms identical to those of Figs. 102 *b* and *c*. Whichever method is used, the switch rating is proportional to the control range $V_{Lb} - V_{La}$; there are thus good reasons for economy and, to minimise distortion, for keeping the control range as narrow as the application will allow.

6.4.2 A.C. regulator circuits: single-phase

The single-phase total regulator circuits are so straightforward that, with resistive load, no explanation of their operation is required. Any of the variations of Fig. 100 can be connected between an a.c. source and a load to provide on–off control (static contactor), i.h.c. control or delay-angle control. The circuit is the same for all; only the control arrangements differ.

An inductive (lagging power factor) load introduces the problem of dV/dt, since the load inductance maintains forward current into the reverse voltage halfcycle; when the load current falls below the holding current, a rapid voltage rise occurs across the thyristor(s) or triac. The single wafer of the triac, handling both directions of current, has a decaying number of current carriers still present at the end of each current halfcycle. A high rate of rise of voltage at this instant can thus more readily induce refiring of the triac and loss of control. A capacitor across the triac, with some series resistance to limit di/dt, successfully controls dV/dt for an inductive load supplied from mains frequencies. The thyristor of a reverse-parallel pair having a different wafer for each direction of current is better placed to withstand the dV/dt stress, having been nonconducting for a period beforehand, and is accordingly better suited to high-frequency applications.

The control characteristic for a resistive load is simple to calculate from the well defined output voltage waveform for each value of firing delay α. Where the load is inductive (lagging power factor) and when the power factor is variable, the output voltage waveform is not determined solely by α; a range of output voltages are obtained for each value of α depending on the power factor (Shepherd, 1965, 1966). The continuation of current in the same direction for a period after the end of the voltage halfcycle renders the firing of the reverse current path of no effect until the forward current has collapsed to

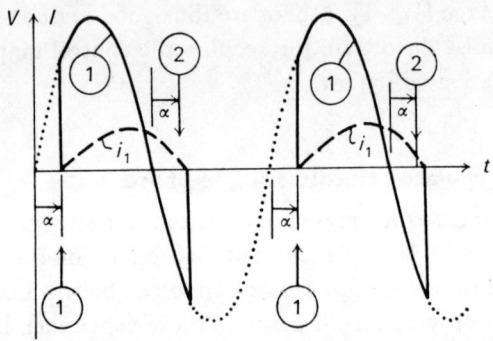

Fig. 103   Pulse-fired a.c. regulator in which the true alternating behaviour cannot develop since the firing pulses to thyristor 2 occur while 1 is still conducting into an inductive load

zero. Firing the switch with short pulses can thus produce asymmetrical behaviour as, for example, in Fig. 103, where one thyristor never conducts because it is always fired while the other is conducting. Prolonged firing pulses lasting 180° are thus preferable to ensure that a true alternating behaviour develops in the steady state. Transiently, the two halves of the switch may for low-power-factor loads conduct asymmetrically, but symmetry develops as the d.c. transient decays.

The way in which the behaviour depends on power factor is shown in Fig. 104, which gives two extreme current waveforms for unity- and zero-power-factor loads. If the current waveform falls to zero before firing the next halfcycle, the load current is discontinuous and

144

the output voltage is less than the supply voltage. No decaying transient can occur since each halfcycle begins from zero load current. If, however, the load current decays to zero after firing the next halfcycle, the switch is permanently closed (in spite of the delay α), and the load voltage equals the supply voltage. A decaying transient component of direct current will now exist, its value depending on the firing angle of the first halfcycle.

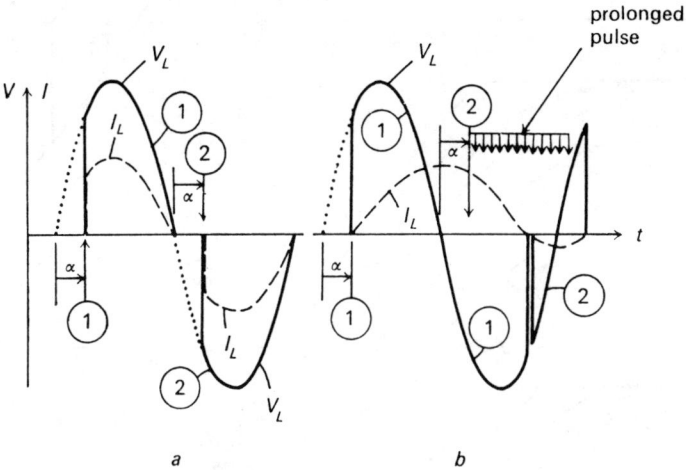

Fig. 104 Output voltage and current for the same delay angle α when the load is (a) resistive and (b) inductive (zero power factor)

Output voltage of the latter equals the supply voltage in spite of the delay α

The ratio of output to input voltage for various load power factors against the delay angle α is shown in Fig. 105. Ratios of r.m.s. and mean values are both given, the latter being useful where the resultant change of flux in a load is to be calculated.

A single-phase a.c. regulator has also been used (Pollard *et al.* 1969) to soft-start a 20 MW supply-frequency induction furnace with parallel capacitors for power-factor correction.

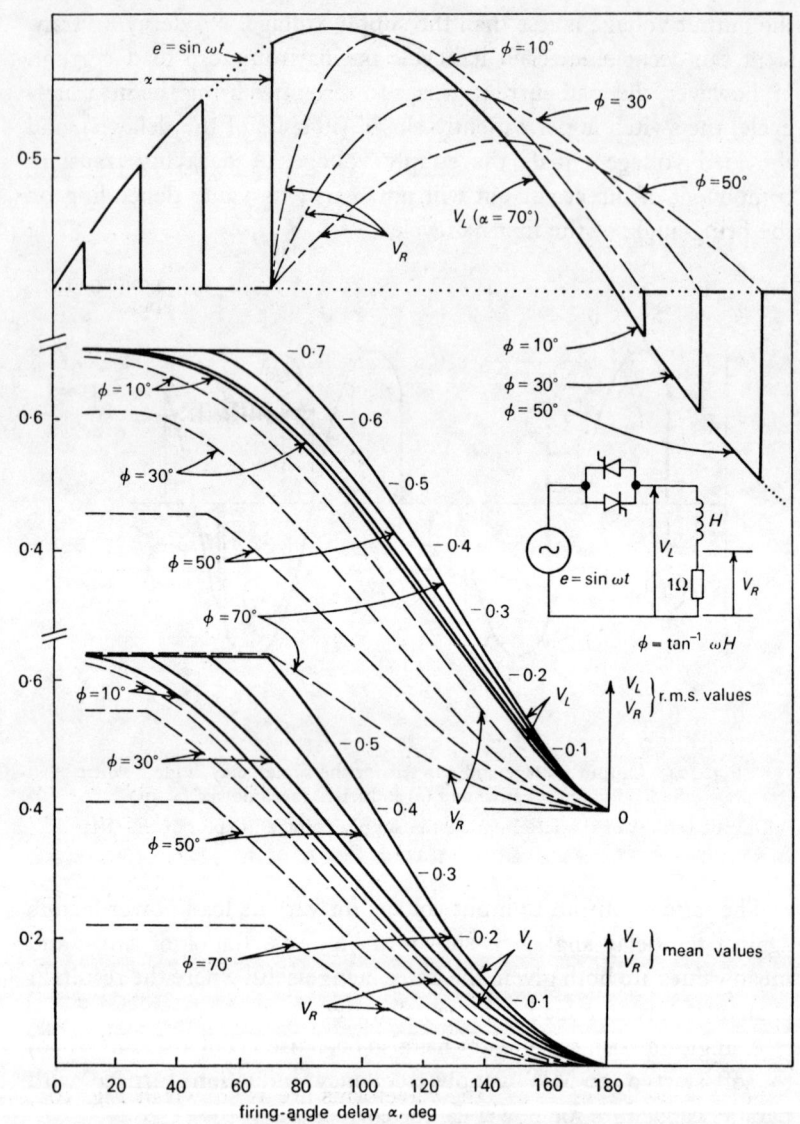

Fig. 105 Control characteristics for an a.c. regulator for inductive loads with various power-factor angles ϕ

At the top, voltage and current waveforms are shown for $\alpha = 70°$; below, curves show the variation of r.m.s. and mean voltages with firing angle. For $\alpha < \phi$, the regulator conducts continuously

6.4.3 A.C. regulator circuits: 3-phase

A greater variety of 3-phase circuits is possible, ranging from the obvious arrangement of one a.c. switch (reverse-parallel thyristor pair or a triac) in each line to circuits requiring only three thyristors. Three examples are shown in Figs. 106 a, b and c.

In Fig. 106 a, the six thyristors controlling a star-connected resistive load are fired in numerical order. Zero delay angle corresponds to the firing of 1 at the beginning of the positive halfcycle of phase a. For

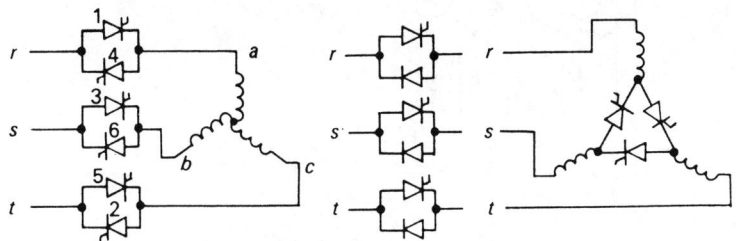

Fig. 106   Three versions of a 3-phase a.c. regulator
Triacs may be used in place of reverse-parallel pairs of thyristors

zero delay, firing 1 initiates a 60° interval in which 5, 6 and 1 conduct together. The thyristor conduction pattern is most easily appreciated for this, and more especially for longer delay angles, by considering the neutral-point potential if only the two fired thyristors connected to the lines with the greatest potential difference were conducting. The case of zero delay is straightforward, since three thyristors conduct throughout each 60° period, so that the load neutral remains permanently at the zero potential of the three supply phases. Each load phase receives the full phase voltage of the supply, giving maximum output.

For a delay angle of 30°, the waveforms are as shown in Fig. 107. When 1 is fired, imagine that only 1 and 6 are conducting, which would produce a load neutral potential following the mean of these phase potentials. It is apparent that phase t is positive with respect to this mean potential for the first 30° and negative with respect to it for the second 30°; hence 5 ceases conduction after 30°. This pattern repeats itself in each successive 60° period, giving the neutral potential as indicated, from which the load phase voltage–time area is readily obtained; it is shown shaded vertically for one halfcycle and horizontally for the other.

For $\alpha > 60°$, only two thyristors conduct together. For $\alpha = 90°$ (Fig. 108) and $\alpha = 120°$ delay, the voltage–time areas shrink as expected, and 150° delay gives the zero output-voltage condition.

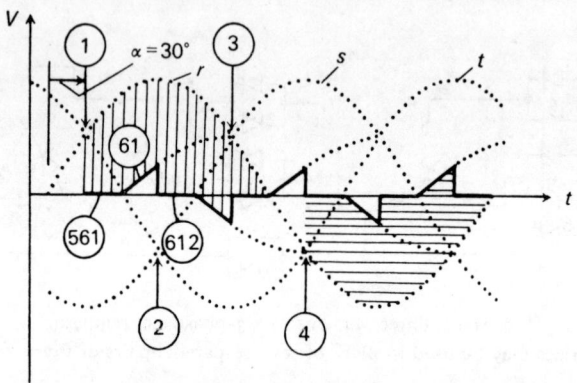

Fig. 107   Voltage–time area (shaded) appearing àcross one phase of
the resistive load for a firing-angle delay $\alpha = 30°$
Solid line indicates the potential of load neutral
with respect to supply neutral

The second arrangement (Fig. 106 b) uses 3 thyristors and 3 diodes. Using the same technique as above, first the neutral potential waveform and then the load phase voltage areas are determined. Load phase voltage–time areas are shown in Fig. 109 for $\alpha = 60°$, 120° and 180°. A delay of 210° is required to reduce the output voltage to zero.

It is emphasised that all the above waveforms are for resistive load only. Thyristor conduction ceases as soon as the line potential becomes less than the load potential, which does not apply to inductive loads, whose current continues to flow beyond the point of voltage equality. The general behaviour with inductive loads (Shepherd,

148

1968) is, however, similar to that of the single-phase a.c. regulator: the output rises with worsening power factor, reaching the maximum output when the phase angle between the end of the voltage halfcycle and the end of the current halfcycle is equal to the delay angle α.

Even when the load is delta-connected, it is necessary to go through the waveform derivations above to establish which thyristors are conducting, after which the line–line output voltage–time areas are readily obtained.

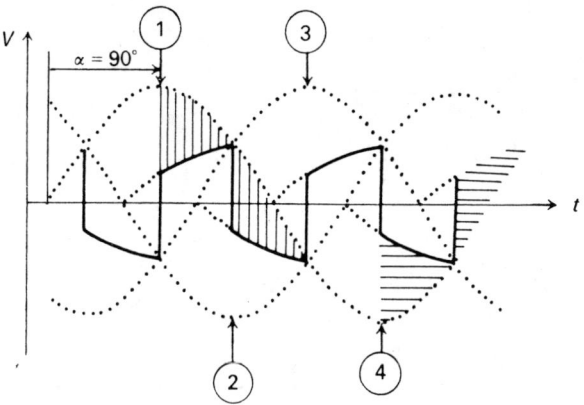

Fig. 108   Voltage–time areas (shaded) across one phase of a resistive load for $\alpha = 90°$

Solid line indicates the potential of load neutral with respect to supply neutral

6.4.4   Load time-constant effects

Although the electrical time constant of an industrial load can be very short, or zero if the load is purely resistive, the relevant process time constant, e.g. temperature in an electric oven, is rarely less than several supply periods, and often orders of magnitude greater. Phase control, though essentially a switching process, produces the effect of smooth proportional control of both the electric power and the process variable. A.C. regulators using i.h.c. control produce low-frequency subharmonics, particularly at the extremes of the control range; process variables with short time constants respond to these subharmonic frequencies. The i.h.c. regulator technique thus falls

149

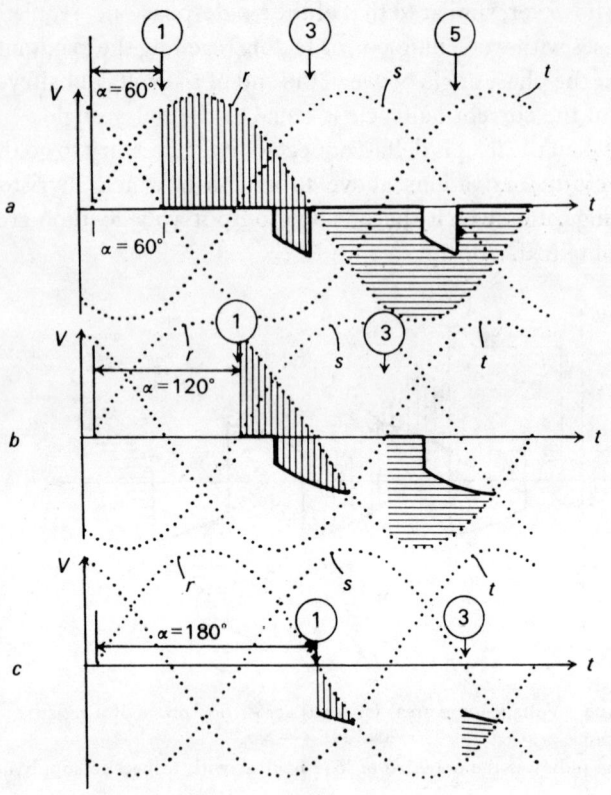

Fig. 109 Voltage–time areas across one phase of a resistive load for a 3-phase a.c. regulator using diode–thyristor pair in each line according to Fig. 106*b*

Delay angles (*a*) 60° (*b*) 120°: (*c*) 180°

between the phase-control technique producing smooth proportional control and the on–off technique for which the on and off periods are of the same order as the process time constant and produce a noticeable variation of the process variable being controlled. The i.h.c.-control technique might be classified as providing pseudoproportional control if the lowest subharmonic period is short compared with the process time constant.

On–off regulators make no attempt to choose a ratio of halfcycles for which the number of halfcycles per repetition period is minimised,

150

as do i.h.c. regulators. The on–off period is determined either manually or automatically using closed-loop control techniques.

Thyristor or triac switches are increasingly displacing the electromagnetic contactor in applications in which the duty cycle renders the reliability or maintenance requirement of the latter unacceptable (Feltbower, 1969).

Motor controls and resistance welding are the most important applications at present for static contactors. For motor controls, the thyristors or triac must be rated for the motor starting current, and the allowable rating is a function of the duty cycle. The thyristor contactor is considerably larger than its electromagnetic equivalent because of the cooling fins.

7 CAPACITOR TURNOFF: CHOPPER REGULATORS AND SERIES INVERTORS

7.2 Introduction

Chapters 3–6 have been concerned with the theory and analysis of naturally commutated thyristor circuits. In every case, there has been a mercury-arc forerunner of each thyristor circuit, and the theory of operation of these circuits dates back to their development in the 1920s and 1930s. The present treatment differs from the historical only in that the method of voltage–time area has been universally applied—a method which is also used for the force-commutated circuits to be considered in this and the next Chapter. It is natural that thyristors should be applied first where mercury-arc devices had previously proved valuable. They have been used in large numbers to date in the circuits analysed in Chapters 3–6; this justifies the full treatment given to rectifiers, synchronous invertors, cycloconvertors and a.c. regulators.

A major growth area for thyristors is in force-commutated circuits, for which there is little or no mercury-arc technology on which to build. The long turnoff times of mercury-arc devices made the use of capacitors for turnoff purposes uneconomical, although one or two applications have appeared during the last 30 years, notably the inversion into a resonant circuit for induction heating. Thyristor turnoff times of typically 20–50 μs make capacitors attractive for turnoff, opening up a wide field for applications using d.c. supplies. Choppers (mark/space regulators) and invertors fall into this category.

The theory of parallel capacitor turnoff is presented first. The thyristor to be turned off has a reverse voltage, previously stored in a capacitor, switched across it often by another thyristor, so that the current previously carried by the thyristor now flows as a discharging current through the capacitor. A series resonant circuit in parallel with the thyristor performs a similar function when the resonant current, opposing the load current during one halfcycle of resonant oscillation, reduces the load current to zero, transferring it to its own capacitor which provides the reverse voltage across the thyristor as before. Despite many circuit variations, the capacitor is effectively in parallel with the thyristor at turnoff, reverse-biasing it and accepting the load current in its stead.

The turnoff process itself is concerned only with obtaining the adequate reverse voltage–time across the thyristor, and carrying its current while it is reverse-biased. Inductive loads, however, provide more prolonged turnoff problems, on account of their stored energy. Either the turnoff capacitor must be enlarged to absorb this energy itself (load-related parallel capacitor turnoff) or an alternative path must be provided for the inductive load current. When the load current is unidirectional, as with choppers, a freewheeling diode forms an ideal alternative path. Invertors, having an alternating load current, require more complex return-current diode paths to provide for both directions of load current, and the treatment of return-current diodes is therefore delayed until the Chapter on invertors.

As choppers apply capacitor turnoff most directly, several chopper circuits are examined. A chopper allows a load to be supplied with a variable direct (but pulsating) voltage from a constant-voltage source, by varying the width of the pulse compared with the intervening off periods. Choppers thus provide efficient and flexible control of direct voltage.

The Chapter is concluded with a totally different turnoff principle: series capacitor turnoff and its particular field of application, the series invertor. The capacitor and thyristor are in series at the instant of extinction.

Series capacitor turnoff uses the thyristor(s) to shock-excite a series resonant circuit. The unidirectional characteristic of the thyristor allows a single unidirectional halfcycle of current to flow, at the end of which the capacitor is charged to a higher voltage than the supply. Thyristor conduction must therefore inevitably cease. Some practical invertors are based on series capacitor commutation, and a bridge sine-wave invertor using series commutation is described.

The further extension of parallel capacitor turnoff to invertors is presented in Chapter 8.

7.2 Theory of parallel capacitor turnoff

For a thyristor to be turned off correctly, the following must take place (Gentry *et al.* 1964):

(*a*) The thyristor current must be reduced to zero.

(*b*) A reverse voltage must appear across the thyristor for a time longer than its turnoff time.

(*c*) The subsequent reapplication of forward voltage must be at a rate less than its dV/dt rating.

(*d*) An alternative path must be provided for the load current where inductive.

The precharged capacitor performs action (*a*) by diverting the load current from the thyristor to itself. It performs actions (*a*), (*b*) and (*c*) by being connected effectively in parallel with the thyristor, so that its stored voltage reverse-biases the thyristor; it has sufficient capacitance to carry the load current for longer than the turnoff time before becoming discharged, and it likewise limits the rate of rise of forward voltage during recharging (Payne and Reeves, 1963). The way in which the circuit performs action (*d*) introduces a fundamentally important distinction between two forms of parallel capacitor turnoff for inductive loads.

If, after thyristor turnoff, the load current is transferred away from the capacitor to a constant-voltage circuit before the load current decays to zero, the circuit employs 'impulse commutation', a fast turnoff process, which is thyristor-related (Bedford and Hoft, 1964). The commutating, or turnoff, capacitor is relatively small, governed as it is by the thyristor turnoff time, and it carries the load current for a relatively short period, typically 50–100 μs. Sections 7.2.1–7.2.4 and 7.3 are devoted to various aspects of impulse commutation.

If the turnoff capacitor carries the load current unaided until the load current has been reduced to zero, the capacitor is much larger in order to absorb all the load energy for a tolerable rise of voltage. The reverse-voltage time thus greatly exceeds the turnoff time, and the commutation process is very much slower; it is load-related since a relationship exists between stored load energy and capacitor energy.

Section 7.2.5 deals with load-related parallel capacitor commutation, and Sections 7.4 and 7.5 deal with series capacitor turnoff.

The distinction between thyristor-related impulse capacitor turnoff and load-related capacitor turnoff arises only for inductive loads. A resistive load does not store energy; therefore, the provision of the alternative path (d) above is unnecessary. The circuit can be equally viewed as using a thyristor-related or load-related turnoff technique. The capacitor has in either case the lower value of the thyristor-related technique.

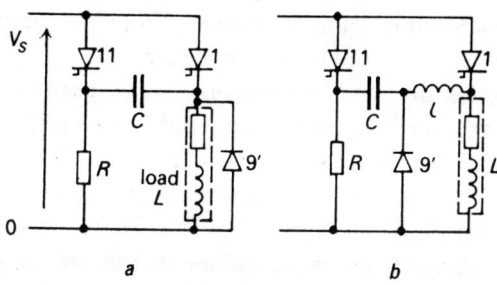

Fig. 110 Principle of parallel capacitor turnoff in which firing 11 places the voltage stored in C as a reverse voltage across 1; inductor l is introduced in (b) to control di/dt at turnon

7.2.1 Switched capacitor turnoff

The precharged capacitor is switched directly across the thyristor 1 to be turned off by a second thyristor 11 in the basic circuit of Fig. 110a. Thyristor 1 conducts the load current keeping the right-hand plate of C at the potential $+V_s$. The left-hand plate potential is reduced to zero as the capacitor charges through R. If, when C is fully charged, thyristor 11 is fired, the left-hand plate of C will be raised to $+V_s$, and hence its right-hand plate to a potential $+2V_s$. The cathode of 1 is thus more positive than its anode; 1 is reverse-biased, and hence ceases conduction. The load current now flows from the supply through 11 and C, discharging C. While C discharges from V_s to zero, 1 is reverse-biased; this is called the reverse-voltage time t_{RV}, which must be greater than the turnoff time of 1. Thus

$$C = \frac{I_L t_{RV}}{V_s} \quad \text{where} \quad t_{RV} > t_{off} \qquad (7.1)$$

155

Eqn. 7.1 is based on a constant load current (= capacitor current) during the period t_{RV}, resulting in a linear discharge of C, thus simplifying the general integral equation

$$C = \frac{1}{V_s} \int_0^{t_{RV}} i_L \, dt \qquad (7.2)$$

During the reverse-voltage time, while 1 has been reverse-biased, the load L supports $2V_s$ initially, decreasing to V_s when C is discharged. The load current must increase somewhat during the reverse-voltage time, owing to the additional voltage across it. After C has been discharged, the load current continues to flow through it, charging it in the reverse direction until the load voltage is zero. From the firing of 11 until the load voltage falls to zero, the path 11–C has been the alternative path for the load current. From this time onwards, it is usual for the load current to transfer to a diode freewheeling path 9' in parallel with the load. Without this, the energy stored in the load would continue to charge the capacitor to a voltage in excess of the economic rating for 1.

An essential feature in any capacitor turnoff circuit is the re-charging of the capacitor ready for the next turnoff process for 1. In the basic circuit of Fig. 110a, this is performed by firing 1 again. A reverse process turns off 11, after which C discharges and recharges to the original polarity.

The above basic technique is used in most chopper and invertor circuits, and is sometimes referred to as impulse commutation (Bedford and Hoft, 1964).

In practice, two thyristors and a capacitor should not be connected in a loop as they are in Fig. 110a, because each will experience high di/dt when fired into the reverse (carrier-storage) conduction of the other. Likewise, when 1 is fired while 9' is conducting a freewheeling current, 1 will experience high di/dt while 9' conducts transiently in the reverse direction. Inductor(s), e.g. l in Fig. 110b, must be included in these two paths to limit di/dt and their presence introduces a measure of resonant behaviour based on the principles of Section 7.2.2.

7.2.2 Resonant capacitor turnoff (thyristor-related)

Resonant turnoff does not require a second thyristor. Instead, the firing of the load thyristor initiates a resonant reversal of the capacitor voltage, followed automatically by a thyristor current reduction and the application of a reverse voltage to the thyristor. Fig. 111 illustrates the basic circuit in which l, C and 1 form the resonant circuit, and losses are neglected (Gentry *et al.* 1964; Bedford and Hoft, 1964).

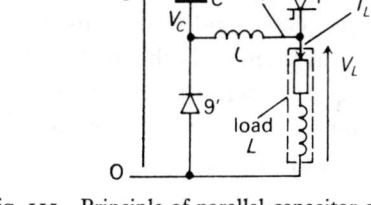

Fig. 111   Principle of parallel-capacitor commutation as it occurs in resonant turnoff circuit

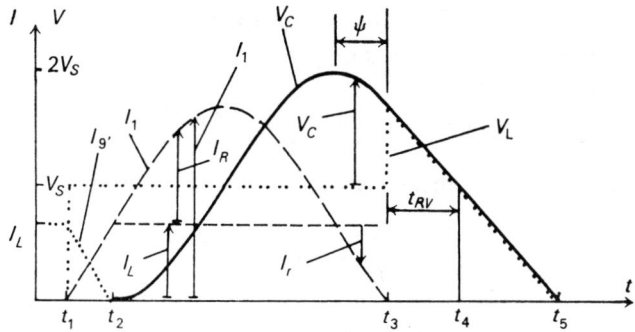

Fig. 112   Voltage and current for circuit of Fig. 111 when thyristor 1 is fired at t_1
Waveforms neglect resistive losses and assume a smooth load current
Eqns. 7.3–7.8 are derived from this Figure

Initially 1 is off and C is charged through L and l with its lower plate at zero potential. When 1 is fired at t_1, any load current which has been freewheeling through 9' and l decays as the current in 1 builds up, their sum being the load current (assumed constant). When all the load current has transferred to 1, 9' ceases conduction at t_2 (Fig. 112).

The resonant circuit lC (and thyristor 1) carries a resonant current I_R charging the lower plate of C positively until it reaches a potential $+2V_s$ after a halfcycle. The next halfcycle of resonant oscillation now begins, during which the resonant current opposes the load current in 1. Provided that the peak resonant current exceeds the load current, the thyristor current is reduced to zero at t_3.

At t_3, the constant load current, now flowing through C and l produces no voltage drop across l so that the remaining voltage V_{c3} ($\pm$) reverse-biases 1. The reverse voltage across 1 persists until t_4 when the load current has linearly discharged C, and $V_{lc} = V_s$, after which C recharges with its lower plate negative until its voltage reaches V_s, whereupon 9' conducts and freewheels the load current. The waveforms illustrating the above description are given in Fig. 112, in which the resonant current I_R has a raised zero at the level I_L, and the capacitor voltage has a raised zero at $+V_s$.

A complete cycle is divided into four intervals. The equations below apply in the first three active intervals, while the fourth is a quiescent interval awaiting the reactivation of the circuit by another thyristor-firing pulse. The following equations arise directly from Fig. 112 using the fundamental equations:

$$di/dt = E/H \quad \text{and} \quad dV/dt = I/C$$

Interval 1 $(t_2 - t_1)$—a linear increase of thyristor current:

$$\frac{I_L}{t_2 - t_1} = \frac{V_s}{l} \tag{7.3}$$

Interval 2 $(t_3 - t_2)$—a resonant reversal of capacitor voltage and reduction of thyristor current to zero:

$$\psi = \sin^{-1}\frac{I_L}{I_{Rp}} \tag{7.4}$$

$$t_3 - t_2 = \frac{180 + \psi}{180}\,\pi\sqrt{(lC)} \tag{7.5}$$

Interval 3 $(t_5 - t_3)$—a linear discharge and recharge of the capacitor

158

providing the reverse-voltage time $t_{RV}(= t_4 - t_3)$:

$$V_{C2} = V_s \cos\psi \tag{7.6}$$

$$\frac{V_{C2}}{t_{RV}} = \frac{V_s \cos\psi}{t_4 - t_3} = \frac{I_L}{C} \tag{7.7}$$

$$t_5 - t_3 = \frac{1 + \cos\psi}{\cos\psi} t_{RV} \tag{7.8}$$

7.2.3 Choppers based on Figs. 110 and 111

While the above circuits have been used to introduce the principles of forced commutation, they each form the basis for a practical

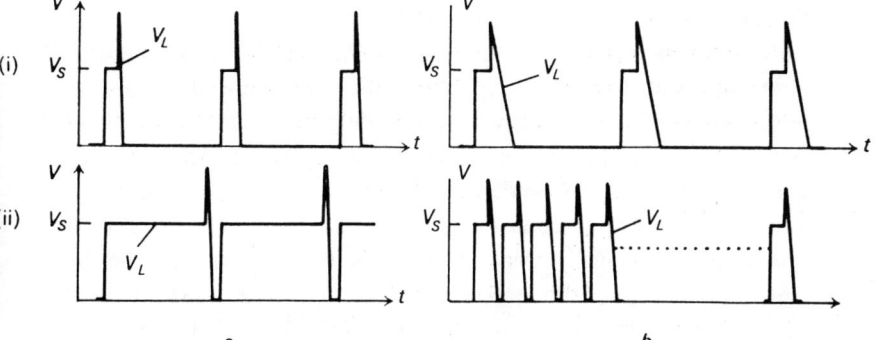

Fig. 113   Output voltage at low output voltage (i) and high output voltage (ii) for chopper circuits based (a) on Fig. 110 and (b) on Fig. 111

Waveforms in (a) have a constant frequency whereas those in (b) require a high frequency for a high output voltage

chopper circuit, the design of which uses the equations given. The 2-thyristor circuit (Fig. 110b) allows on and off periods for the load to be separately controlled, so that mark/space-ratio control at a constant frequency is possible, as shown in Fig. 113a. The resonant circuit has a relatively constant on time determined primarily by the resonant period, so that mark/space-ratio control is only possible if the firing rate for the thyristor is varied (Fig. 113b).

7.2.4 Thyristor voltage during t_{RV}

In both the basic circuits above, a substantial, but decreasing, reverse voltage has appeared across the thyristor to be turned off. This is beneficial in sweeping carriers out of the junction nearest the thyristor anode, but it does little to hasten the recombination of carriers in the centre junction. The turnoff time, determined primarily by carrier lifetime in the centre-junction region, is not greatly influenced by the magnitude of the reverse voltage during t_{RV}.

The sudden rise of reverse voltage at the beginning of t_{RV} can have troublesome effects elsewhere in a circuit, where it may appear as a high dV/dt across another thyristor (Rice and Nickels, 1968). For this and other reasons (to be explained in Section 7.5), it is often desirable to connect a diode in reverse parallel with the thyristor to be turned off. If a diode is so connected in Fig. 110a, the capacitor C, on firing 11, is rapidly discharged through this diode, drastically shortening t_{RV}. If, however, C has sufficient series inductance, as in Fig. 110b or Fig. 111, the reverse-parallel diode does not shorten t_{RV} but lengthens it. The circuit behaviour is explained with reference to Fig. 114, in which the added diode plays no part until the thyristor reverse voltage appears at t_3. Now the diode 4′ prevents this, and allows the resonant current to continue flowing through the diode until t_4, when the diode ceases conduction (Fig. 114). At t_4, the resonant current again equals the load current, so that, with neither the diode nor the thyristor conducting, the remainder of the re-charging process for C is linear. In this case, the added diode lengthens t_{RV} to nearly double, thus allowing the use of a smaller capacitance, given by

$$t_{RV} = t_4 - t_3 = \pi\sqrt{(lC)}\frac{90 - \psi}{90} \qquad (7.9)$$

7.2.5 Load-related capacitor turnoff

Both Figs. 110 and 111 have used thyristor-related (impulse) commutation, since the load current has been diverted from the capacitor, when charged, to a diode freewheeling path. The reverse-voltage time is only marginally greater than the turnoff time at maximum-load-current conditions.

Load-related capacitor turnoff does not transfer the load current from the capacitor to a diode path, but instead the capacitor is enlarged to accept the load current until it has decreased to zero, for an

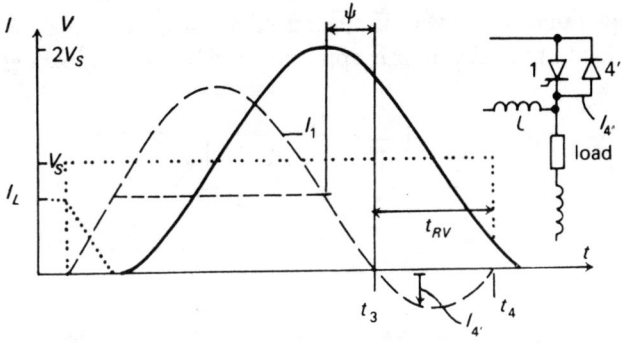

Fig. 114 Alterations to waveforms of Fig. 112 when thyristor is fitted with reverse-parallel diode 4', allowing the resonant current to continue flowing between t_3 and t_4

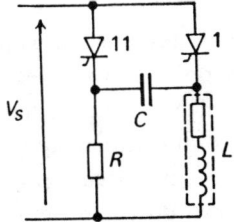

Fig. 115 Parallel capacitor turnoff using load-related technique

Since the freewheeling diode is absent, the capacitor, appropriately enlarged, must absorb the entire load energy after turnoff

acceptable rise of capacitor voltage. In a lossfree circuit, the capacitor must accept the energy associated with the reduction of the load current to zero. Although Fig. 115 is not a desirable alternative to Fig. 110 with its freewheeling diode, it does illustrate load-related capacitor turnoff in its simplest terms. (Applications of load-related capacitor turnoff with invertors are discussed in Section 7.5 and Chapter 8.)

Fig. 115 has no freewheeling diode. When 11 is fired, 1 is reverse-biased as before with the initial voltage on the capacitor ($= V_s$). It is assumed that, when the load current, now flowing through 11 and C, is reduced to zero, the forward voltage across 1 rises to V_{1x} ($> V_s$). A rigorous analysis of the damped resonant transfer of energy from the load to the capacitor naturally involves the solution of the differential equations. However, if the resistive losses are small (high Q-factor load), the behaviour is approximated by the lossfree behaviour.

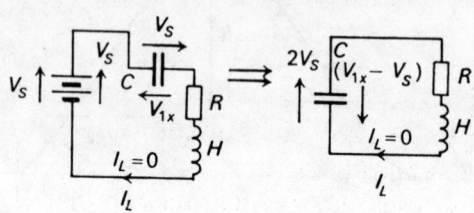

Fig. 116   Equivalent circuits for Fig. 115 which apply
following the firing of thyristor 11
Initial conditions are shown outside, and final
conditions inside, the circuit loop

As the resonant loop includes the d.c. supply, there is an energy contribution from the supply, and it is easier to use an equivalent circuit for Fig. 115, in which the supply has been eliminated and its voltage incorporated into the initial and final capacitor voltages. Fig. 116 derives the equivalent circuit after 11 has been fired. The initial conditions are shown on the diagram outside the circuit and the final conditions inside it. From Fig. 116, we have, neglecting R,

$$\tfrac{1}{2}HI_L^2 + \tfrac{1}{2}C(2V_s)^2 = \tfrac{1}{2}C(V_{1x} - V_s)^2 \tag{7.10}$$

yielding
$$C = \frac{HI_L^2}{V_{1x}^2 - 2V_{1x}V_s - 3V_s^2} \tag{7.11}$$

which has a positive denominator only if

$$V_{1x} > 3V_s \tag{7.12}$$

Consider briefly a comparison of the capacitor required for load-related and thyristor-related capacitor turnoff. Let the load inductance be 100 mH carrying 10 A from a 100 V supply (implying a modest load time constant of 10 ms, if the resistance is not neglected). If $V_{1x} = 4V_s$, the capacitance calculated from eqn. 7.11 is 200 μF. For a thyristor turnoff time of 20 μs, the capacitance required, from eqn. 7.1, is 2 μF.

The above example emphasises the benefit of using impulse commutation, for which a diode path relieves the turnoff capacitor of the load current and also clamps the circuit voltages so that lower thyristor voltage ratings can be used. The penalty, usually outweighed, is the switching, rather than resonant behaviour of the impulse turnoff system, resulting in output voltage waveforms which are rectangular rather than sinusoidal. Sine-wave invertors, however, require a large capacitance for waveshaping; it matters little whether it is used for load-related parallel capacitor turnoff or for filtering a rectangular waveshape. A practical application for load-related turnoff therefore exists for sine-wave invertors, provided that the changes in the load are small enough to give consistent resonant behaviour and acceptable output-voltage regulation.

Series capacitor turnoff (Section 5.3) is another load-related technique which meets the requirements for sine-wave invertors.

7.3 D.C.–D.C. choppers

The requirements for an ideal chopper are, in order of importance:
(a) a thyristor voltage rating only marginally higher than the supply voltage
(b) a thyristor current rating only marginally higher than the maximum load current
(c) the minimum of turnoff components
(d) a high efficiency.
(a), (b) and (c) keep down the component costs. As thyristors become cheaper, however, (a) and (b) become relatively less important.

All choppers operate by sequentially connecting and disconnecting a d.c. supply and a load, so that the load voltage switches between the supply voltage and (usually) zero voltage. Load-voltage variation is obtained by varying the relative durations of the supply voltage and the zero-voltage intervals. The terms 'time-ratio control' or 'mark/

space-ratio control' are used to describe it. Consequently, the load voltage can be varied from zero to a value approaching the supply voltage (Borst *et al.* 1966).

Load current only flows from the supply during on periods. An inductive load requires a freewheeling path to circulate the load current during the off period. Thus, where the freewheeling diode conducts load current for an appreciable period, the mean load current exceeds the mean supply current. For inductive loads, therefore, the

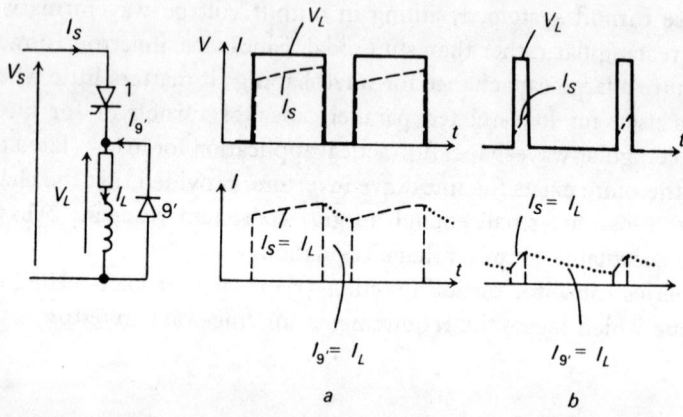

Fig. 117 Voltage and current for chopper controlling finite inductive load

I_S and I_L rise together when the thyristor is on, but, when off, $I_S = 0$ and I_L ($= I_{9'}$) falls exponentially. Waveforms (*a*) and (*b*) apply to high and low output, respectively

chopper acts like a d.c. variable transformer, reducing the output voltage and supply current approximately in proportion. The above ideas are summarised in Fig. 117. For practical examples, see Beasley and White, 1965; Jaquet *et al.* 1969; Bailey and Varley, 1969; Davis, 1969; Band and Stephens, 1969; Kalis and Lemmrich, 1969.

Choppers can also be designed to give a controllable output voltage which exceeds the supply voltage. The principle is to energise the load from the supply, or from the supply and an energy store connected in series for voltage addition, whichever is higher. While the load is energised from the supply alone, the energy store is charged from the supply also, increasing the total supply current. When fully charged, the energy store is arranged to give up its energy to the load circuit, raising the load voltage above the supply voltage. The stepup chopper is rare compared with the stepdown chopper.

Chopper circuits almost universally use parallel capacitor turnoff, either switched or resonant, or some combination of these.

7.3.1 Basic 2-thyristor chopper

The variety of chopper circuits is such that only a few can be considered here. As stated in Section 7.2.3, the basic circuit arrangements of Fig. 110*b* with its minor addition of an inductor to control *di/dt* form a practical chopper circuit of the utmost simplicity. Fig. 110*b* is reproduced here as Fig. 118. The two thyristors 1 and 11 conduct alternately. When 1 is fired, the capacitor *C* turns 11 off and vice versa. Variable output is obtained by varying the mark/space ratio as indicated in Fig. 113*a*. The positioning of the *di/dt* control inductor *l* is interesting, as it must protect each thyristor from high *di/dt* when firing into the other,

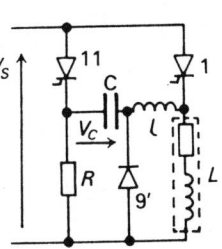

Fig. 118 Practical 2-thyristor chopper circuit

and it must also protect the load thyristor when firing into the freewheeling diode 9'. It is also desirable that the inductor should not be in series with the load circuit, so that its current rating can be minimised. The position shown meets all these requirements. The turnoff thyristor 11 never fires into the conducting freewheeling diode, as either 1 or 11 is always conducting. The inductor *l* can be linear (usually air-cored) or on a saturating iron core. The latter is arranged to support the supply voltage for about 50 μs before saturation occurs, by which time a large proportion of the thyristor wafer will be conducting. Thereafter the inductor presents only its saturated inductance to the circuit.

Two limiting conditions apply for most choppers: maximum load current which determines the turnoff capacitor, and correct recharging of the capacitor which determines minimum output.

For Fig. 118, maximum output is defined as refiring the load thyristor 1 when the capacitor has become fully charged. For this condition, the load current, assumed smooth, is a maximum I_{Lx}. At $t = 0$ in Fig. 119, thyristor 1 has been conducting, so that C is charged to V_s through 1, l, C and R. Thyristor 11 is fired at t_1, causing a resonant current to rise through 11, C and l and in reverse through 1. For practical values of di/dt, l is small, so that the peak resonant current exceeds the load current many times. A linear rise

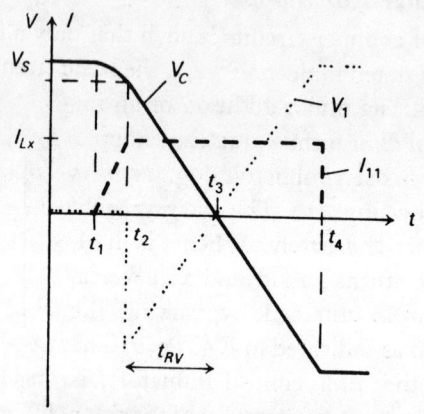

Fig. 119   Voltage and current for the circuit of Fig. 118
during the turnoff period for 1

Eqns. 7.13–7.20 are based on these waveforms

of current is therefore a good approximation to the actual sinusoidal rise. Assuming that i'_{11} is the rate of rise of current in 11 which provides the desired factor of safety on the di/dt rating of the thyristor, we have

$$l = V_s/i'_{11} \qquad (7.13)$$

If the capacitor current rises to equal the load current, at time t_2, the capacitor becomes slightly discharged to V_{C2} ($< V_s$), given by

$$V_{C2} = V_s - \frac{I_{Lx}(t_2 - t_1)}{2C} \qquad (7.14)$$

since the average current during this period is $I_{Lx}/2$.
Also

$$t_2 - t_1 = I_{Lx}/i'_{11} \qquad (7.15)$$

166

The above equations give V_{C2} in terms of known factors and C. The capacitance can now be calculated from the reverse-voltage time t_{RV} chosen to give the required factor of safety over the turnoff time. No voltage drop across l occurs since the capacitor current is now constant, giving a linear discharge, so that

$$\frac{V_{C2}}{t_{RV}} = \frac{I_{Lx}}{C} \tag{7.16}$$

giving, from eqns. 7.14, 7.15 and 7.16,

$$C = \frac{I_{Lx}}{V_s}\left(t_{RV} + \frac{I_{Lx}}{2i'_{11}}\right) \tag{7.17}$$

Thyristor 1 becomes forward-biased as the capacitor recharges with reverse polarity. When the right-hand plate becomes negative with respect to the supply voltage zero, the freewheeling diode begins conduction and clamps the capacitor voltage at this level. The linear recharge time $t_4 - t_3$ is given by

$$t_4 - t_3 = t_{RV}\frac{V_s}{V_{C2}} \tag{7.18}$$

so that the on period for thyristor 11, assuming 1 is refired at t_4, is $t_4 - t_1$, given by

$$t_4 - t_1 = \frac{I_{Lx}}{i'_{11}} + t_{RV}\left(1 + \frac{V_s}{V_{C2}}\right) \tag{7.19}$$

The r.m.s. current rating at maximum chopper output for thyristor 11 is readily calculated from eqn. 7.19:

$$I_{11r} = I_{Lx}\sqrt{\left(\frac{t_4 - t_1}{T}\right)} \tag{7.20}$$

in which T is the repetition period of the chopper; it is assumed that $t_2 - t_1 \ll t_4 - t_1$.

The load voltage waveform at maximum output, which derives from Fig. 119, is shown in Fig. 120. The corresponding voltage–time area VTA_L for a repetition period T is seen to be

$$VTA_{Lx} = V_sT + \tfrac{1}{2}V_{C2}t_{RV} - \tfrac{1}{2}V_st_{RV}\frac{V_s}{V_{C2}} \tag{7.21}$$

giving a maximum load voltage V_{Lx}

$$V_{Lx} = V_s - \frac{t_{RV}}{2T}\left(\frac{V_s^2}{V_{C2}} - V_{C2}\right) \tag{7.22}$$

167

Eqn. 7.22 shows that the difference between the maximum load voltage and the supply voltage increases at higher frequencies, at which the reverse-voltage time becomes a greater proportion of the repetition period. The loss of capacitor voltage during the initial capacitor-current rise also reduces the maximum output. Consider now the minimum-output condition, which is governed by how quickly C charges through 1 and R during a short on period for 1.

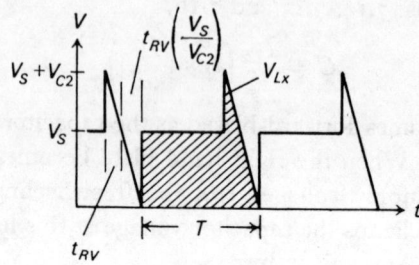

Fig. 120 Load voltage V_{Lx} and voltage–time area (shaded) at maximum output
Eqns 7.21 and 7.22 are based on this Figure

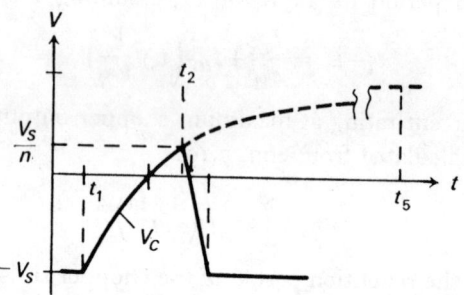

Fig. 121 Exponential recharging voltage waveform of capacitor after 1 is fired at t_1, when the chopper is operating at its minimum output voltage $V_{Ln} \simeq V_s/n$

If a minimum output current $I_{Ln} = I_{Lx}/n$ is required, the resistor R for the steady-state condition can be calculated as follows, with reference to Fig. 121, in which 1 is fired at t_1. The current previously flowing through 11 and R is small compared to the load current, so that 11 is reverse-biased with the full capacitor voltage $V_{C1} (= V_s)$ when 1 is fired. The capacitor discharges and charges exponentially

168

through R, l and 1, since the voltage drop across l can be ignored for the small rate of change of current involved. As the steady-state load current is I_{Lx}/n, the capacitor need only be charged to V_s/n to provide sufficient reverse-voltage time for 1. Thyristor 11 can thus be refired at t_2 (Fig. 121), so that the minimum on time for 1 is $t_2 - t_1$, given by

$$t_2 - t_1 = CR \log_e \frac{2n}{n-1} \tag{7.23}$$

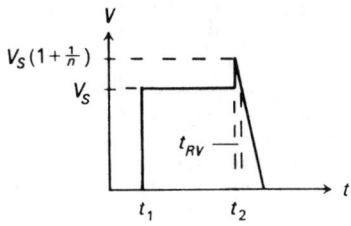

Fig. 122 Load voltage and voltage–time area at minimum output as defined for steady-state conditions when the capacitor need only be charged to V_s/n for correct turnoff

The waveform of the load voltage (Fig. 122) can now be calculated. The minimum load voltage–time area VTA_{Ln} is

$$VTA_{Ln} = V_s(t_2 - t_1) + \frac{V_s}{2}\left(1 + \frac{1}{n}\right)(n+1)\, t_{RV} \tag{7.24}$$

giving $\qquad V_{Ln} = \dfrac{V_s}{n} = \dfrac{V_s}{T}\left\{ CR \log_e \dfrac{2n}{n-1} + \dfrac{(n+1)^2}{2n}\, t_{RV} \right\}$ $\qquad$ (7.25)

in which T is the repetition period.

Note that the minimum output calculated above applies only to the steady-state or slowly changing outputs. If the chopper, in a closed-loop control system, is suddenly called on to operate at minimum output after previous operation at maximum output, the circuit must turn off maximum load current at minimum load voltage. The capacitor must therefore be fully precharged to V_s, and not merely to V_s/n. The same is true for back-e.m.f. (motor) loads where the circuit is required to operate at full current and minimum output voltage simultaneously.

For the capacitor voltage to be charged to within 5% of the supply voltage, a charging time $t_5 - t_1 = 3\cdot7CR$ is required (Fig. 121). The output-voltage waveform, shown in Fig. 123, is also based on Fig. 121, but with an on time for 1 of $t_5 - t_1 = 3\cdot7CR$. Whereas transiently I_L

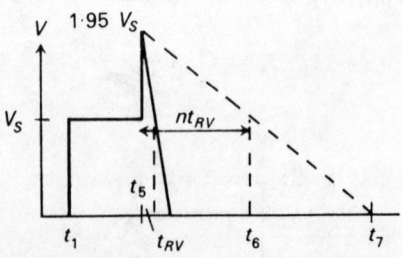

Fig. 123   Minimum output voltage and voltage–time area when the load current is I_{Lx} (solid) and I_{Ln} (dashed), showing the inevitably increased reverse-voltage time nt_{RV} and load voltage–time area VTA_{Ln}

will result in a reverse-voltage time t_{RV}, the steady-state minimum output must be calculated for the steady-state load current I_{Ln} ($= I_{Lx}/n$) giving a reverse-voltage time of approximately nt_{RV}:

$$VTA_{Ln} = V_s(t_5 - t_1) + \frac{1\cdot95 V_s}{2} \frac{1\cdot95 nt_{RV}}{0\cdot95} \tag{7.26}$$

$$= V_s\left(3\cdot7CR + \frac{1\cdot95^2 nt_{RV}}{1\cdot9}\right) \tag{7.27}$$

so that
$$V_{Ln} = \frac{V_s}{n} = \frac{V_s}{T}\left(3\cdot7CR + \frac{1\cdot95^2 nt_{RV}}{1\cdot9}\right) \tag{7.28}$$

Eqn. 7.28 allows R to be calculated from the other known terms. If the chosen repetition frequency is too high, a negative R results. Even a moderately low frequency may give an R which is not substantially greater than the load resistance—a situation which makes for poor efficiency at reduced output. Such disadvantages as these have led to the development of improved circuits, which, however, are analysed using the same techniques as above.

The current rating for thyristor 11 at maximum chopper output is almost entirely determined by the turnoff process, the resistor R providing generally only a small additional current for the short on time of 11, as indicated in eqn. 7.20. At minimum steady-state output, thyristor 11 is conducting the resistor current V_s/R for much of the repetition period. The turnoff process still adds somewhat to the r.m.s. current in 11, giving approximately

$$I_{11r} = \sqrt{\left\{\left(\frac{V_s}{R}\right)^2 + I_{Ln}^2 \frac{t_7 - t_5}{T}\right\}} \qquad (7.29)$$

Eqn. 7.20 or 7.29, whichever gives the higher value for I_{11r}, is used to decide its r.m.s. current rating.

The basic 2-thyristor chopper is usually operated with a constant repetition period, and the on time is varied. This is not a necessary restriction, however; and, if the load ripple current at 50 % output is too great, the repetition period can be shortened to keep the peak-to-peak ripple below a set value. Lower-frequency operation is allowable near maximum and minimum output, ensuring that they are kept, respectively, as high and as low as possible.

7.3.2 Resonant turnoff chopper

The principles of Section 7.2.2 underlie all resonant turnoff choppers. The basic circuit of Fig. 111 also serves as a practical chopper circuit (McMurray, 1963) which is defined by eqns. 7.3–7.8. No further circuit description or analysis is necessary, but it is useful to derive equations which summarise the operation of this resonant chopper in terms of just two factors, the angle of turnoff ψ (Fig. 112) and the reverse-voltage time t_{RV}. Working from eqns. 7.3–7.8, we have the following time intervals:

$$t_0 - t_1 = lI_L/V_0 \qquad (7.30)$$

in which, from the equality of resonant energies,

$$l = CV_s^2 \Big/ \left(\frac{I_L}{\sin\psi}\right)^2 \qquad (7.31)$$

and, from eqns. 7.6 and 7.7,

$$C = I_L t_{RV}/V_s \cos\psi \qquad (7.32)$$

171

giving, on substitution in eqn. 7.30,

$$t_2 - t_1 = \frac{t_{RV} \sin^2 \psi}{\cos \psi} \tag{7.33}$$

and, on substituting for C and l in eqn. 7.5,

$$t_3 - t_2 = \frac{180 + \psi}{180} \pi t_{RV} \tan \psi \tag{7.34}$$

Eqn. 7.8 gives
$$t_5 - t_3 = \frac{1 + \cos \psi}{\cos \psi} (t_{RV})$$

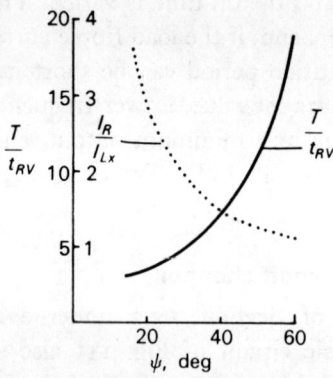

Fig. 124 Variations with turnoff angle ψ of the design parameters T/t_{RV} and I_R/I_{Lx} for the resonant turnoff chopper of Fig. 111

The shortest period ($= 1/f_x$ where f_x is the maximum frequency corresponding to maximum output) is given by

$$\frac{1}{f_x} = \frac{t_{RV}}{\cos \psi} \left\{ \sin^2 \psi + \frac{180 + \psi}{180} (\pi \sin \psi) + 1 + \cos \psi \right\} \tag{7.35}$$

which, with eqns. 7.31 and 7.32 for l and C in terms of V_s, L_L, ψ and t_{RV}, fully defines the circuit. Typical values for ψ lie between 20° and 60°, for which f_x and the ratio I_R/I_{Lx} take the values shown in Fig. 124.

There is no restriction on minimum output, as the thyristor may be fired as infrequently as desired. At very low frequencies, however, the load current will cease to be continuous, and will become a train of current pulses.

In the interests of economy, the capacitor required for Fig. 111 is kept as small as the turnoff time of the thyristor allows. The requirement that the peak resonant current exceeds the maximum load current places a maximum value on the resonant inductor l. The resonant frequency is therefore largely predetermined by the turnoff time giving a value for f_x of several kilohertz, for typical turnoff times. If it is desirable that f_x should be lower, a saturating resonant inductor can be used. The peak resonant current, and turnoff behaviour, are governed by its saturated inductance, which has the desirable low value, but the on time is based largely on its unsaturated behaviour.

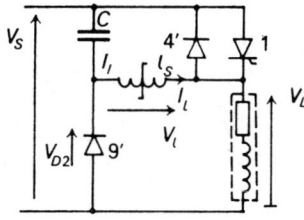

Fig. 125 Resonant turnoff chopper using a saturating resonant inductor l (Morgan circuit) to extend the on time without increasing C

Much longer on periods are therefore obtainable, reducing the maximum frequency.

The Morgan circuit (Morgan, 1963) uses a saturating resonant inductor. It is worth consideration for its own sake and because saturating inductors find wide application in thyristor circuits.

The Morgan circuit is given in Fig. 125 in a slightly modified form. The diode 9′ has been moved to its present position from directly across the load. This alteration avoids a high di/dt in the thyristor when fired into the conducting freewheeling diode.

Initially thyristor 1 is off, and the load current freewheels through 9′ and l_s saturated (referred to as reverse saturation). Thyristor 1 is fired at t_1 (Fig. 126), applying V_s to the load and placing a forward voltage across l_s. First, the current in l_s is reduced to zero while the

173

core remains saturated. Secondly, the applied voltage drives the core towards forward saturation. While the flux is changing in l_s, its magnetising current is neglected, so that it is considered as an open circuit. No change in the capacitor voltage can therefore occur until l_s reaches forward saturation at t_2. The usual forward resonant current now flows through 1, l_s and C, resonantly reversing the capacitor voltage, which is completed at t_3. C now places a voltage across l_s, which drives the core towards reverse saturation, during which change

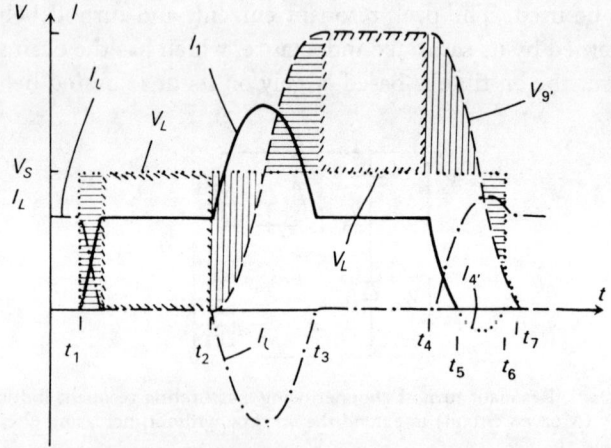

Fig. 126   Waveforms for circuit of Fig. 125 in which the rectangular boundary-shaded voltage–time areas are spent in driving the core flux from positive to negative saturation, and vice versa; the vertically and horizontally shaded areas appear across l_s in its saturated state and sum to zero

l_s again behaves as an open circuit. At t_4, l_s reaches reverse saturation, so that the resonant halfcycle of the resonant current now begins to flow in the reverse direction through 1 reducing its current to zero at t_5. The excess of the resonant current over the load current flows through the diode $4'$ until t_6, when resonant current again equals the load current. The recharging of the capacitor to V_s now takes place at constant current I_L until freewheeling action restarts at t_7 completing the cycle.

7.3.3 Further chopper developments

Fig. 126 shows how quiescent periods are introduced into the cycle by the unsaturated behaviour of the saturating resonant inductor. A quiescent period can also be introduced by inserting a reverse-parallel thyristor–diode pair in the resonant loop using a non-saturating inductor. Fig. 127 illustrates the idea. Firing 1 no longer starts the resonant action; it only energises the load. The capacitor, charged to V_s at the previous turnoff process, remains with its lower

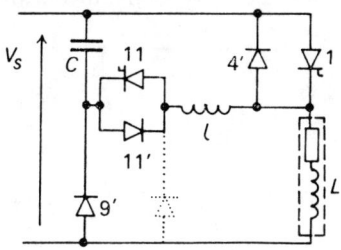

Fig. 127   Extension of the on time in a resonant turnoff chopper by delaying the start of the resonant current until 11 is fired

plate near zero potential, so that, when it is desired to turn off 1, the firing of 11 will initiate the resonant reversal, the next halfcycle of which is completed through 11′ and 4′, reverse-biasing 1. Thus the on time for 1 can be chosen or varied at will, the circuit possessing the same control flexibility as Fig. 110. Diode 4′ may be omitted, and 9′ may be repositioned as shown dotted, relieving 11′ of the free-wheeling current.

Having moved 9′ to the dotted position, the relative positions of C and 11/11′ are immaterial. The diode–thyristor pair can be reversed, the thyristor conducting downwards (Fig. 128). The resonance now starts through diode 11′ immediately 1 is fired, adding to the current in 1 early in its conduction period when the current has not spread evenly to the whole wafer; the reverse resonance to perform turnoff is delayed at will until 11 is fired. For the polarity of 11 and 11′ shown, C is not automatically charged through the load when the supply voltage is applied, so that 11 must be fired once specifically for this purpose. Another resonant inductor l_{11} can be inserted in series with 11′ as indicated by the arrow; this reduces the amplitude and increases the period of the preparatory resonant reversal, a considerable advantage to thyristor 1; but 1 must have an on period always longer than the (increased) reversal period. Fig. 128, or a variation on it, is a popular chopper circuit which has found application in voltage control of d.c. traction motors.

All choppers discussed so far have transferred the load current to a circuit which includes the turnoff capacitor and the supply; i.e. the capacitor has been effectively in parallel with the thyristor to be

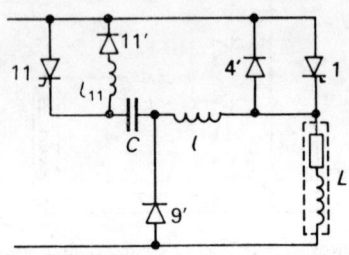

Fig. 128   Modification of Fig. 127 in which turnoff occurs ψ degrees after firing 11, rather than $180 + \psi$, as is the case for Fig. 127

Additional resonant coil l_{11} may be added to reduce the amplitude of the resonant current which flows immediately 1 is fired

turned off. As all d.c. supplies to choppers have a considerable capacitance added in parallel, it is equally possible to divert the load current from both the thyristor and the supply (Fig. 129). On firing 1, C is charged resonantly to $2V_s$ through 11′. Any time later, 1 can be turned off by firing 11. Although less obviously, this circuit still uses parallel capacitor turnoff owing to the presence of C_s. The only difference between Figs. 127 and 129 is that the latter has the resonant circuit returned to the zero-voltage supply terminal instead of the $+ V_s$ terminal.

7.3.4   Stepup choppers

The principle of the stepup chopper is the storage of energy at the low supply voltage and its release at a higher output voltage. The thyristors merely act as switches essential for the transfer of the energy storage component from the input circuit to the output circuit.

The most convenient energy storage component is the inductor since, neglecting losses, it can be charged (in time) from any voltage (however low) and will discharge into a high-voltage load. A capacitor in parallel with the output is required to smooth the output voltage, since the energy is supplied in pulses (Borst *et al.* 1966).

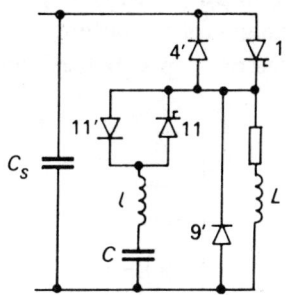

Fig. 129 Resonant turnoff chopper in which the turnoff is delayed until 11 is fired

The mechanism remains that of parallel capacitor turnoff owing to the presence of C_s

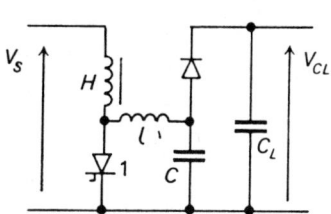

Fig. 130 Stepup chopper using the inductor H to store energy from V_s and release it at V_{C0}

Thyristor uses simple resonant turnoff

A stepup chopper can make use of any turnoff mechanism for the thyristor. A simple one based on the resonant turnoff circuit is shown in Fig. 130. Thyristor 1 is fired to charge H with current from the supply V_s; it may be convenient, if l is a saturating inductor, to give a more prolonged on period for 1 and hence a longer charging time. When l and C turn 1 off, the current flowing in H and l first recharges C; and when $V_c = V_{CL}$, both capacitors rise in voltage together, absorbing energy from H. The output voltage rises until the load consumes the energy transferred. Voltage regulation is consequently poor, and a feedback regulator would be essential for a varying load to stabilise the output voltage. A sensing circuit which lowers the frequency of a unijunction transistor pulse-rate generator as the voltage exceeds the desired value would provide suitable control. Typical circuits are given in Chapter 9. The variable-frequency chopper makes smoothing more difficult, and increases the likelihood of an inconvenient beat effect in the load, should it contain an oscillator or similar electronic circuits.

A modification of Fig. 130 which is useful for a high voltage ratio is strongly reminiscent of the ignition coil for a petrol engine. The inductor is tapped as shown in Fig. 131.

A high voltage ratio means that little is gained from the auto-transformer principle; a double-wound inductor (transformer) can be used to provide useful isolation for the high-voltage output. It is now apparent that the transformer-core material is poorly used and

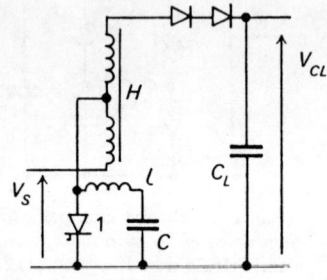
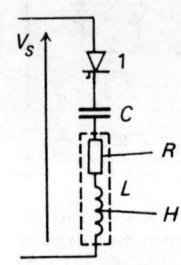

Fig. 131 Stepup chopper which avoids the need to rate 1 for the output voltage V_{C0}

Fig. 132 Basic series-capacitor turnoff circuit

that the flux density changes covering less than half of the total available excursion of $2B_x$. This is a shortcoming of all stepup chopper circuits. A more efficient and compact arrangement would therefore have an invertor and a rectifier. Although it uses a few more components, it has a far superior voltage regulation. The square-wave output, when rectified, requires little smoothing. It is likely that an invertor–rectifier arrangement would be used in preference to a step-up chopper for more than a few kilowatts.

7.4 Theory of series capacitor turnoff

The thyristor and capacitor are in series for this turnoff technique (Fig. 132). The thyristor can only conduct transiently while the capacitor is charging, and when the capacitor is fully charged, turnoff is inevitable, since the capacitor leakage current is less than the holding current of the thyristor. With an inductive-load resonant capacitor, charging takes place so that the capacitor voltage when fully charged exceeds the voltage of the charging supply; the thyristor is thus usefully reverse-biased, further assisting turnoff. The thyristor in Fig. 132 can conduct only one pulse of current; for repeated operation, the capacitor must be discharged after each charging process. This is done either through the load giving a reverse-direction load current pulse, or separately, in which case the load carries unidirectional current pulses only. The two possibilities are shown

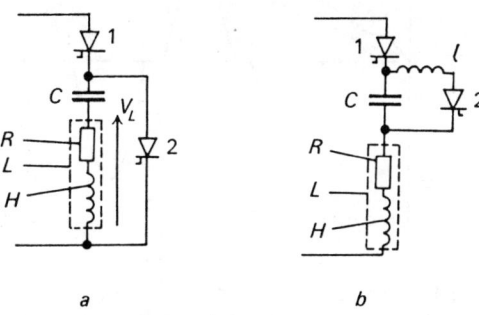

a   b

Fig. 133   Additional thyristor 2 added to Fig. 132, which discharges the capacitor and allows a repeated charge–discharge sequence to be obtained with alternate firing of 1 and 2

In circuit (a), the capacitor discharge current flows through the load, and in (b) it does not

in Figs. 133 a and b. In Fig. 133 a, the two opposing halfcycles of load current have equal durations, so that the circuit operates as an invertor—a series-capacitor-commutated invertor. In Fig. 133 b, the unidirectional pulses of current through the load can be made frequent or infrequent by the choice of firing rate, giving a high or low mean value to the direct load current. The series capacitor turnoff technique for a chopper regulator based on Fig. 133 b is inferior to the parallel capacitor turnoff technique, and is thus not discussed further. Series-capacitor-commutated invertors, however, are in common use for sinusoidal outputs, using many variations of Fig. 133 a (Bedford and Hoft, 1964; Mapham, 1967 b). This basic invertor circuit is therefore

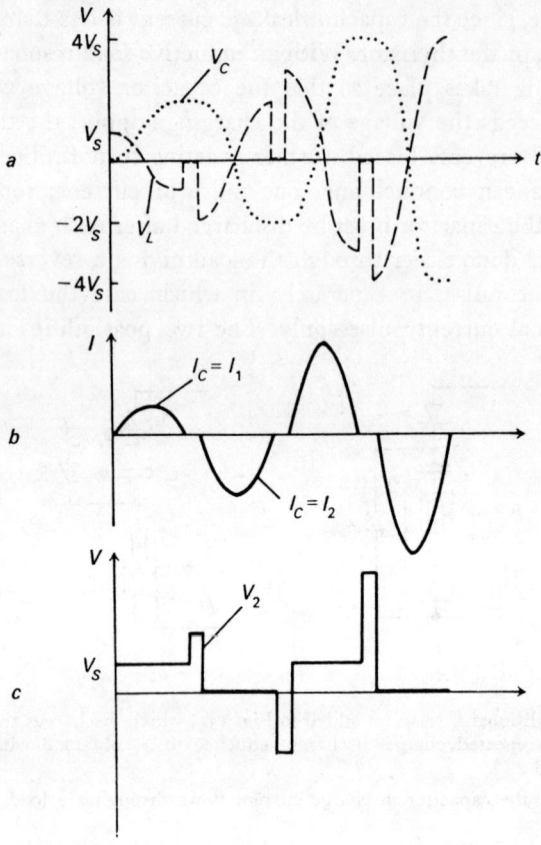

Fig. 134 Voltage and current for Fig. 133 a in which the load resistance is
zero, giving a progressive rise of amplitude with each succeeding halfcycle

analysed briefly to yield the basic equations which underlie all series
invertors. The alternate firing of the two thyristors of Fig. 133 a
progressively increases the amplitude of the capacitor and load voltages, as shown in Fig. 134, which illustrates these waveforms for the
extreme case of zero load resistance. Load voltage and load current
are nearly 90° out of phase, the distortion in the voltage waveform
accounting for the difference. The power factor differs from zero
only to the extent that the energy content of the resonant circuit is
rising so that the input power cannot be zero.

When load resistance is present, the resonant circuit is damped, so that the voltage buildup reaches a ceiling. The current wave does not lag by as much as before, as the current waveform has its peak earlier, rising more steeply and decaying more slowly.

The waveforms of Fig. 134 exaggerate the time required for turn-off; in practice, the waveform V_2 across the HCR circuit can be considered rectangular with a direct voltage component of $V_s/2$ and a rectangular alternating voltage of amplitude $V_s/2$. A short turnoff time also implies that the frequency of this square wave is almost equal to the resonant frequency of the HCR circuit. At the resonant frequency, the inductor and capacitor combined behave as zero resistance, so that the fundamental component of the square wave appears wholly across the resistive component of the load. Hence

$$V_{Rp} = \frac{V_s}{2}\left(\frac{4}{\pi}\right) \tag{7.36}$$

and
$$I_{Rp} = V_{Rp}/R \tag{7.37}$$

so that V_{Hp}, V_{Cp} and V_{Lp} can be readily calculated with acceptable accuracy, noting that all values are peak in eqns. 7.36 and 7.37. The capacitor supports a direct voltage component $V_s/2$ in addition to the alternating component. Harmonic voltages also appear across the load, more so than across the capacitor, as is clear from Fig. 134a, which shows voltage steps of amplitude V_s every halfcycle. In the steady state, the loss of voltage during a halfcycle of resonant oscillation due to the damping effect of the resistance is exactly made up by the step of voltage above.

The harmonic voltages appearing across the load are calculated from the amplitude of the harmonic present in the square wave applied to the HCR circuit and the attenuating effect of the circuit elements at the harmonic frequency in question. Where the HR components form the load, high harmonics will be present in the load voltage, owing to the low impedance of the capacitor to high frequencies. A purer load-voltage waveform can be obtained by connecting the load in parallel with C and using a separate component for H. A later variation of the basic circuit shows how the capacitor and load may be connected so that the capacitor and load do not experience the component of the direct current which appears in this simple circuit.

By firing the thyristors at a frequency lower than resonance, the invertor has an improved voltage regulation (Kusko and Szpakowski, 1965). The current waveform has longer zero periods than shown (exaggerated) in Fig. 134 b, so that the applied voltage wave can no longer be considered rectangular. To calculate the output voltage waveform, a harmonic analysis of the applied voltage wave is required; its components are then used as before.

7.5 Series-capacitor-commutated invertors: development

The basic circuit of Fig. 133 a when used as an invertor has the disadvantages of drawing current from the supply only once per cycle when once per halfcycle is preferable, and of high harmonics in the load voltage when the load inductance is used to form the resonant circuit. Various circuit improvements are described below which overcome these and other disadvantages and allow a greater operating flexibility (Thompson, 1963).

First, the invertor can be arranged to draw power from the d.c. supply on both halfcycles by connecting the lower end of the resonating load to the midpoint of the d.c. supply, where available, or to an 'artificial' midpoint where not, as shown in Fig. 135. If the capacitance of C_s is large, the midpoint potential is stiff and all the capacitor-voltage variations occur across C_1. However, for a low impedance supply, C_s can be reduced to $C/2$ and C short-circuited, in which case all the capacitor-voltage variations occur across the capacitors C_s, which now act together as C.

Secondly, where $C_s \gg C$, the load may be connected in parallel with C, rather than in series, provided a separate inductor h is inserted in series with C, to give the correct resonant behaviour (Fig. 136). The two capacitors C_s ensure that no direct voltage component can appear across C or across the load, while C, providing a low impedance for the higher harmonics, ensures a smoother load-voltage waveform. The parallel-connected load allows operation at zero load ($R = \infty$) when the damping is zero, whereas the series-connected load allows operation at short-circuit load ($R = $ o) but not at zero load.

Thirdly, reverse-parallel diodes may be added to the thyristors of any previous circuit. The resonant action now continues after the first halfcycle, the second halfcycle flowing through the diode and discharging the capacitor from its peak value. The reverse voltage appearing across the thyristor after the first halfcycle is now limited

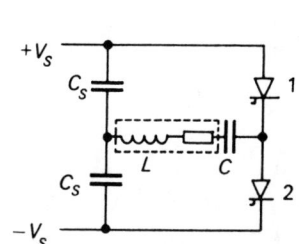

Fig. 135 Modification of the basic circuit so that direct current is taken from the supply during each halfcycle of output

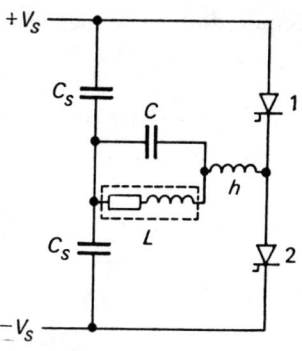

Fig. 136 Improvement of output voltage waveform by connecting the load across C and introducing a separate resonant inductor h

to the forward-voltage drop in the diode. With a firing frequency lower than resonance, the continuation of resonant behaviour through the diode during the period preceding the next thyristor-firing pulse gives a voltage wave applied to the resonant circuit, which is square, and not stepped as indicated in Fig. 134c for the basic circuit. Harmonic content is reduced, and the voltage regulation with load changes is improved.

Fourthly, the circuit is readily converted to a bridge configuration (with or without diodes) by replacing the two capacitors C_s of Fig. 136 by thyristors. The voltages applied to the resonant circuit are so doubled and the capacitors C_s eliminated. Most high-power invertors (> 5 kVA) use the bridge configuration.

All the above developments except the addition of diodes do not involve a change to the equivalent circuit, and hence the behaviour is identical to the basic circuit of Section 7.4. The behaviour of an invertor with diodes is described later in this Section.

Fifthly, the inductor h of Fig. 136 can be duplicated and placed in series with each thyristor, as shown in Fig. 137 (Mapham, 1967a). With no mutual coupling, the operation is unaffected, except that the inductor now has to operate with unidirectional current pulses, so that its B/H loop is poorly utilised and an air-gapped core will be required. Where the windings of h_1 and h_2 are perfectly coupled,

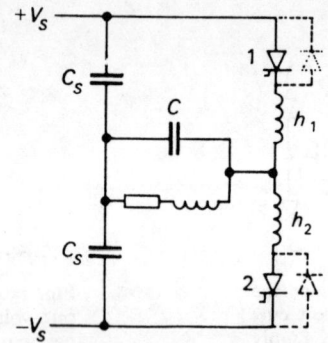

Fig. 137 Replacement of h in Fig. 136 by h_1 and h_2 which, when mutually coupled, allows a new mode of operation at a frequency higher than resonance using parallel-capacitor commutation

however, a new mode of operation is possible. For firing frequencies below resonance, the series capacitor turnoff mechanism continues to operate. However, it is now possible to use a firing frequency higher than resonance, which implies that the incoming thyristor is fired before the resonant action has turned the other off. Assume that 2 has been conducting and the right-hand plate of C has already become more negative than the potential $-V_s$. Firing 1 will place a voltage greater than $2V_s$ across h_1 and, by transformer action, the same voltage will be induced across h_2, reverse-biasing 2 and turning it off. As will be appreciated later, the circuit is now using parallel commutation. The difference between series and parallel commutation no longer involves physical circuit differences, but is merely a function of how the thyristors are fired. 2-mode invertor operation is thus possible, e.g. series-commutated sine-wave operation for normal loads, and parallel commutation for short-circuit faults, say. Diodes cannot be fitted where mutual coupling exists between h_1 and h_2, since the diode prevents the correct resonant voltages from being developed across the inductor windings.

7.5.1 Bridge sine-wave invertor

Finally, a bridge sine-wave invertor is described, using series commutation and diodes. The circuit is shown in Fig. 138, in which 'snubbing' networks for dV/dt and di/dt suppression have been

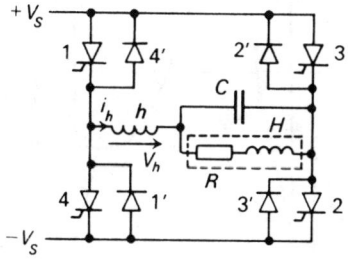

Fig. 138   Basic circuit for practical sine-wave bridge invertor

omitted. This suppression is required since, when 1 is fired, it inevitably places a high dV/dt across 4 etc. (Rice and Nickels, 1968). The description begins at $t = 0$, with the capacitor charged to $2V_s$ (negative or positive), a resonant current at its peak value flowing through $1'$, h, C and $2'$ and a load current flowing through H and R. The voltage across h is thus zero, giving no rate of change of current in h. At $t = 0$, 1 and 2 are fired, and they take over immediately the current in $1'$ and $2'$, which cease conduction. The polarity of C is such that $4V_s$ now appears across h, raising the current in it from its previous value at a corresponding rate. The potential V_{Ch} is initially $-3V_s$, so that an undamped resonant reversal about the potential $+V_s$ would give a final potential of $+5V_s$. However, in the steady state, the damping must be such that the final potential V_{Ch} is $3V_s$,

185

giving the damped resonant reversal shown (dashed) in Fig. 139. When the current in thyristors 1 and 2 has decreased to zero, the capacitor is charged with V_{Ch} at a potential $+3V_s$, so that the reverse halfcycle of oscillation begins through 3′ and 4′, until the voltage across h is again zero, at the maximum diode current. At this instant, thyristors 3 and 4 are fired to initiate the next invertor halfcycle. Although the inductor current waveform is noticeably nonsinusoidal,

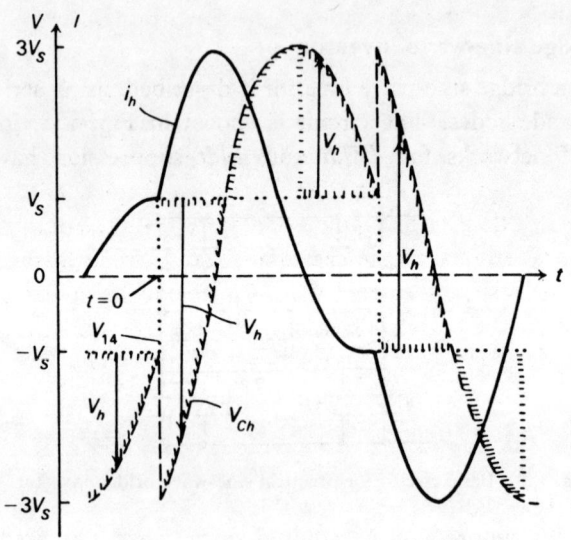

Fig. 139   Voltage and current for Fig. 138

Potentials are drawn with respect to a zero at the midpoint of the d.c. supply. V_{14} and V_{Ch} symbolise the potentials at the points in the circuit common to devices 1 and 4, C and h

the capacitor voltage waveform bears a close resemblance to a sine wave. The capacitor voltage waveform V_C is obtained for Fig. 139 as

$$V_C = V_{Ch} - V_{32} \qquad (7.38)$$

and, since $V_{32} = -V_{14}$ (they are antiphase square waves),

$$V_C = V_{Ch} + V_{14} \qquad (7.39)$$

Both V_{Ch} and V_{14} waveforms appear in Fig. 139, and their addition to give an approximate sine wave is easily visualised.

186

It is apparent that the series commutation allows the diagonally opposite device paths of Fig. 138 to produce a square-wave voltage across h, C, H and R. The components h and C form a lowpass filter feeding the load. Although the operation of the circuit has been described in terms of series capacitor turnoff, the same output voltage waveform would be obtained by applying the Fourier components of the square wave to the filter and load in turn, and summing the output voltage components. The similarity between a series-commutated sine-wave invertor and a parallel-commutated square-wave invertor with an output filter is thus demonstrated.

The capacitor can be considered as being made up of two parallel components, one to resonate the load, and render the inductive load purely resistive, and the other to perform the series capacitor turnoff function. It is then clear that a lagging power-factor load requires a larger capacitor than a unity-power-factor load. As the peak resonant current is proportional to the capacitance, and circuit losses depend largely on the resonant current which flows equally at the full load or a light load, the circuit losses of an invertor designed to operate over a wide load and power-factor range will be much greater than those for an invertor feeding a fixed unity-power-factor load. A typical efficiency for this type of circuit at the 10 kVA level is 85 %.

8 PARALLEL-CAPACITOR-COMMUTATED INVERTORS

8.1 Introduction

Invertors have exactly the same circuit configurations as rectifiers, namely centre-tap and bridge, single-phase and polyphase. In a rectifier, the alternating voltages produce a fixed conduction sequence for the rectifier devices; whereas, in a capacitor-commutated invertor, the conduction sequence of the thyristor is fixed by the firing-control circuits, and the sequential device conduction generates the alternating-voltage output. In a power rectifier, the alternating-voltage waveform is that of the supply, i.e. sinusoidal, and the direct-voltage output includes a ripple component, made up of parts of the sine waves. The inductance usually present in the load gives smooth direct current and a rectangular alternating current. For a capacitor-commutated invertor, the roles of voltage and current are reversed. The direct voltage is smooth by virtue of the capacitance connected across the d.c. supply to assist commutation; hence, as the thyristors are merely switches, the alternating output voltage is basically rectangular. The direct current includes a ripple component, which is a part of the alternating-current waveform. A sinusoidal output voltage is obtained by filtering, using inductors and capacitors in resonance where the resonant energy is large compared with the load power per halfcycle, giving limited damping, and hence a good approximation to a sinusoidal waveshape.

Centre-tap invertors, like rectifiers, require a transformer, since the windings energised by the thyristors carry the unidirectional thyristor currents. The transformer converts the several unidirectional m.m.f.s into an alternating m.m.f., thus producing the alternating output voltage. The direct current flows through each thyristor and its winding in turn, the d.c. supply being connected to the transformer centre tap (Fig. 140). The number of output phases is q, where

188

q is odd, and $q/2$ where q is even. The mechanism by which one thyristor in the sequence of conduction is turned off by firing the next (called self commutation) is explained in Section 8.2, in which the alternative arrangement using an auxiliary thyristor specifically for turnoff purposes is also presented.

The principles of centre-tap invertors are introduced during the treatment of the single-phase ($q = 2$) circuit. These principles underlie the operation of both centre-tap and bridge invertors, and the reader should, therefore, consider this material an essential foundation to the understanding of all parallel capacitor-commutated in-

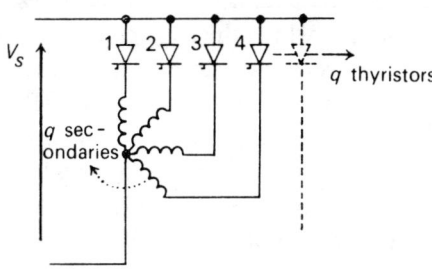

Fig. 140   Generalised centre-tap invertor configuration giving q-phase
output when q is odd, and a $q/2$ phase output when q is even

vertors (Bedford and Haft, 1964; Murphy and Nambiar, 1961). Such vital aspects as operation at varying power factor, load-related capacitor commutation, the introduction of return-current diodes, commutation energy and circulating currents are all discussed in relation to this simple invertor. Turnoff using an auxiliary thyristor is then presented. Finally, the 3-phase centre-tap invertor is considered briefly for an inductive load only, since a power rating requiring a 3-phase invertor will usually be better provided by a bridge circuit.

Bridge invertors, like bridge rectifiers, have the major advantage of not needing a transformer except where required for voltage transformation or for isolation. When an invertor feeds a rotating machine, it is usually possible to choose the motor voltage or d.c.-supply voltage to suit each other, and there is no need for isolation. The equipment is thus smaller and cheaper.

Bridge invertors use the same principles as centre-tap invertors, so that Sections 8.2–8.5 are essential for the understanding of bridge circuits. Their behaviour with resistive and inductive loads is similar to centre-tap invertors; return-current diodes may be simply added in the form of a bridge rectifier, avoiding the current-transfer mechanism between tightly coupled half primaries which is associated with centre-tap invertors. Where an output transformer exists, a tapped primary winding with the return-current diode bridge connected to the taps remains useful in avoiding circulating currents, although other techniques will also be discussed.

Bridge invertors are used for sine-wave and square-wave outputs. Sine-wave outputs can be obtained, either by series (or load-related parallel) commutation, in which the large commutating capacitor and inductor resonate at a frequency close to the invertor frequency, or by filtering the output of a square-wave invertor (Bedford and Hoft, 1964; Ott, 1963). Square-wave invertors almost universally use return-current diodes for their clipping action, and for the ability they impart to the invertor to operate efficiently at varying power factor with little change in the output voltage. Section 7.5 has already developed a series-commutated configuration for a bridge invertor.

As with centre-tap invertors, the turnoff process for the thyristors of a bridge invertor can be performed either by firing the next invertor thyristor in the sequence or by firing an auxiliary thyristor. The basic turnoff techniques are first extended to bridge-invertor circuits, after which the single-phase and 3-phase bridge invertors are described for inductive loads only. For single-phase invertors, a constant direct voltage is tacitly assumed, so that the possible effect of a variable direct voltage does not arise until the 3-phase circuit is treated. This is appropriate, as it is the variable-speed a.c. motor drives which require both 3-phase and variable voltage.

8.2 Capacitor turnoff for centre-tap invertors

Parallel capacitor turnoff is the most commonly used method for invertor circuits, which use many variations of the basic principles described in Section 7.2. It is therefore appropriate to develop two of these before embarking on an analysis of invertor circuits.

A thyristor in a centre-tap invertor can be turned off either by firing the next load-carrying thyristor in the sequence by means of a capacitor connected between the two thyristors, or by firing an auxiliary (non-load-carrying) thyristor, capacitor-coupled in the same way. These two basic techniques are shown for a single-phase centre-tap invertor in Figs. 141 *a* and *b*.

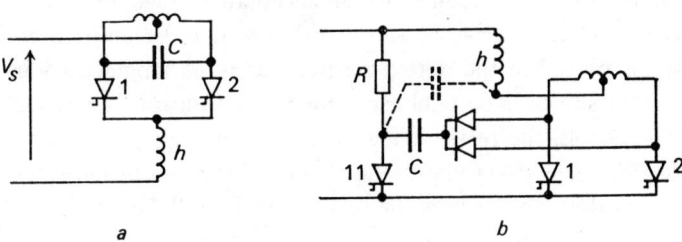

Fig. 141   Single-phase centre-tap invertor ($q = 2$) using
(*a*) self commutation and (*b*) auxiliary commutation

The capacitor in (*b*) may be moved to the dotted position
and the diodes omitted when the load is resistive

8.2.1   Self commutation

Firing 1 in Fig. 141 *a* turns 2 off (and vice versa) since, while 2 is conducting, the right-hand plate of the capacitor is near zero potential, so that, by transformer action, the left-hand plate is at a potential near $2V_s$. When 1 is fired, the capacitor prevents any instantaneous change of potentials at the transformer primary to which it is connected, so that the cathode potential of 1 jumps suddenly to near $2V_s$. As the anode of 2 is likewise held near zero potential, the desired reverse voltage has been obtained. The current which flowed previously through 2 now flows through C and 1 in accordance with the principles of parallel capacitor turnoff.

The purpose of the inductor h is now apparent, since it supports the voltage pulse of amplitude $2V_s$ when 1 is fired. Without it, an excessive current would flow from the supply, demanding a much larger turnoff capacitor. Commutation techniques frequently require the insertion of an inductor, which plays, as it does here, an important role in the turnoff process. This circuit is analysed later in Section 8.3.

8.2.2 Auxiliary commutation

Fig. 141*b* illustrates the idea behind the use of an auxiliary thyristor for capacitor turnoff. The capacitor reverse-biases both load thyristors via the two diodes (although only one needs the reverse voltage at a time). The capacitor must absorb the load energy, and the energy in the inductor h, for an acceptable capacitor voltage rise. When the load is resistive, the two diodes may be omitted and the capacitor placed in the dotted position. It is no longer obvious that the circuit still uses parallel capacitor turnoff until it is realised that, during turnoff, the resistive load generates no voltage (unlike an inductive load), so that the presence of load resistance in series with the thyristors does not destroy the principle of parallel capacitor turnoff. It is also interesting to note that the turnoff capacitor carries 'load' current during turnoff, since the current in the inductor h just before turnoff is a reflection of load current. During turnoff, h supports a voltage pulse of amplitude $2V_s$ as before, which increases the current in h during turnoff, placing some additional load on the turnoff capacitor.

An important advantage of capacitor turnoff with an auxiliary thyristor is the ability to include off periods in the invertor sequence, allowing periods of zero output voltage to be included in the output voltage waveform. The technique can be used for contactless on–off control, for waveform improvement (Turnbull, 1963) and for output-voltage control. The circuit can be viewed as an intimate combination of an invertor and a chopper.

8.3 Single-phase centre-tap invertor
8.3.1 Resistive load

A resistive load poses the fewest problems since there is no stored energy in the load at commutation. The circuit diagram is shown in Fig. 142, in which the transformer half primaries have the same number of turns as the secondary to simplify the analysis. In the simplest mode, a thyristor is fired only when the transient of the preceding commutation has died away; square-wave invertors use this mode. The two thyristors are fired alternately, developing the sum of the supply and inductor voltage alternately across each half primary. Firing one thyristor turns the other off by forcing across it a reverse voltage, stored in C. Assume that 2 is conducting: the right-hand terminal of the primary winding is at zero potential for the above mode since there is no rate of change of current (neglecting transformer magnetising current) to develop a voltage across h. By autotransformer action, the left-hand terminal of the primary is at a potential $2V_s$. The capacitors C_s and C prevent any transformer

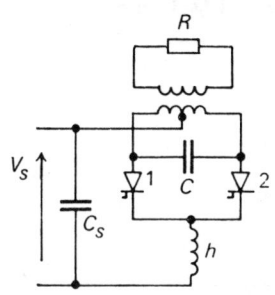

Fig. 142 Basic circuit of single-phase centre-tap invertor feeding a resistive load

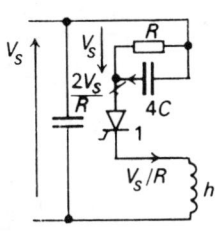

Fig. 143 Equivalent circuit of Fig. 142 immediately after 1 has been fired

Initial conditions are shown inside the circuit loop

terminal from changing its potential suddenly. When 1 is fired, there-fore, the cathode potential of 1 is raised to $2V_s$, so that $2V_s$ now appears across h (hence its purpose), and a reverse voltage $2V_s$ also appears across 2, turning it off. Two actions now occur together: the capacitor discharges and recharges, thus controlling the start of the next halfcycle, and the voltage across h, initially $2V_s$ but falling to zero when the capacitor is fully recharged, builds up a current in h from the supply, this current dividing equally between the two half primaries and thus contributing no m.m.f. The combined action is more easily understood from the equivalent circuit in Fig. 143.

The capacitor has a value $4C$ owing to the $2:1$ turns ratio of the autotransformer; and, when the transient produced by one com-mutation has fully died down before the next thyristor is fired, the voltage across $4C$ at t_1, the firing instant for 1 in Fig. 144, is V_s. The initial current in h for these conditions is V_s/R. These initial conditions are shown inside the circuit loop of Fig. 143, which result in an initial current in $4C$ of $2V_s/R$. In Fig. 144, thyristor 2 should not be fired before t_1' for the above initial condition to be valid.

The equations governing the response after firing 1 are

$$\begin{bmatrix} I_h' \\ V_C' \end{bmatrix} = \begin{bmatrix} 0 & \dfrac{1}{h} \\ -\dfrac{1}{4C} & -\dfrac{1}{4RC} \end{bmatrix} \begin{bmatrix} I_h \\ V_C \end{bmatrix} + \begin{bmatrix} \dfrac{1}{h} \\ 0 \end{bmatrix} V_s \qquad (8.1)$$

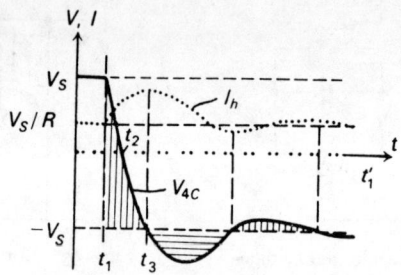

Fig. 144 Voltage across the equivalent capacitor V_{4C}
following the firing of 1 at t_1

Inductor current I_h reaches its maximum value at t_3, the inductor having supported the vertically shaded voltage–time area between t_1 and t_3.

which yield a characteristic equation $\det | s\mathbf{I} - \mathbf{A} | = 0$

of
$$s\left(s + \frac{1}{4RC}\right) + \frac{1}{4hC} = 0 \qquad (8.2)$$

which has complex (oscillatory) roots when

$$\frac{1}{hC} > \frac{1}{16R^2C^2} \quad \text{or} \quad h < 16R^2C \qquad (8.3)$$

For all but the highest values of h, the response of the capacitor voltage V_C (= output voltage) is a damped oscillation (Fig. 144). The natural frequency ω_0 is

$$\omega_0 = \sqrt{\left(\frac{1}{4hC}\right)} \qquad (8.4)$$

The damping ratio ζ is $\quad \zeta = \dfrac{1}{2R}\sqrt{\left(\dfrac{h}{4C}\right)}$ $\qquad$ (8.5)

and the exponential rate of decay α_e of the oscillations is

$$\alpha_e = \frac{1}{8RC} \qquad (8.6)$$

At t_3, the voltage across the inductor has fallen to zero, and the inductor current is a maximum, being made up of its initial current V_s/R, now flowing through R, and the peak buildup (or magnetising) current $I_{hB} = VTA_{h12}/h$, resulting from the shaded triangular voltage–time area across h, flowing through $4C$. That the current I_{hB} splits equally between the two half primaries is apparent from the rate of change of voltage it produces across C. In the equivalent circuit,

$$\frac{dV_{4C}}{dt}\bigg|_{t_3} = \frac{I_{hB}}{4C} \qquad (8.7)$$

and, by autotransformer action,

$$\frac{dV_C}{dt} = 2\left(\frac{dV_{4C}}{dt}\right) \qquad (8.8)$$

Therefore $\qquad\qquad\qquad I_{C3} = \tfrac{1}{2}I_{hB}$ $\qquad\qquad$ (8.9)

The other half of I_{hB} in the real circuit flows through the other half primary, so that no m.m.f. is produced in the transformer by I_{hB}.

For correct thyristor turnoff, the reverse-voltage time $(t_2 - t_1)$ $(= t_{RV})$ must exceed the turnoff time by some safety margin, typically 25 %. If the first part of the capacitor voltage response is approximated by a straight line of the initial slope,

$$4C \simeq 2t_{RV}/R \qquad (8.10)$$

and, using the same approximation, the voltage–time area across h during t_{RV} is

$$VTA_{h12} \simeq 1 \cdot 5 V_s t_{RV} \qquad (8.11)$$

A convenient value for h is one which gives a total current in h at t_2 of $2V_s/R$, thus making the capacitor current at t_1 and t_2 equal and improving the validity of the above assumption about linear discharge. Thus

$$h \simeq \frac{1 \cdot 5 V_s t_{RV}}{V_s/R} = 1 \cdot 5 R t_{RV} \qquad (8.12)$$

which yields a damping ratio $\zeta = 0.433$. If a decay to zero of the transient following commutation is taken to mean a final amplitude 1 % of the initial amplitude, the maximum invertor frequency for the above square-wave operation is

$$f_x = 10^6/37t_{RV} \text{ hertz} \qquad (8.13)$$

The response clearly depends on the initial conditions as well as on eqn. 8.1. By choosing the square-wave operation for which the transient has fully decayed before the next commutation, it is possible to define the initial conditions as the steady-state conditions. If,

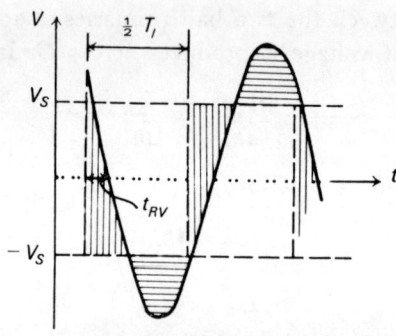

Fig. 145 Capacitor (or output) voltage for a single-phase invertor in which the firing of the next thyristor is advanced to occur within the previous capacitor voltage response, to yield a waveform which is approximately sinusoidal

however, the frequency is raised or the parameters C or h are increased, one commutation begins during the transient following the previous one. Either an analogue-computer simulation or an iterative digital-computer program which matches initial and final conditions can be used to obtain the output-voltage waveform. The waveform of the output voltage can be obtained approximately from the damping ratio and natural frequency above, in conjunction with the concepts expressed in Fig. 145. Here attention is focused on the voltage–time areas appearing across H during the commutation (shaded vertically) and afterwards when h has a reverse voltage across it (shaded horizontally). The vertically shaded area increases the

196

current in h from its initial value, and the horizontally shaded area reduces it. In the steady state, the mean voltage across h must be zero, equalising the shaded areas. The amplitude of the waveform adjusts itself in successive halfcycles until the vertically and horizontally shaded areas in a halfcycle are equal, while retaining the shape set by the natural frequency and damping ratio already found.

For an approximately sinusoidal output waveform (Fig. 145), the half period $\frac{1}{2}T_I$ of the invertor should be about $4t_{RV}$. For a given load resistance and invertor frequency f_I, $t_{RV} = 10^6/8f_I$, and C and h are calculated from eqns. 8.10 and 8.12. The resonant half period is found to be

$$\tfrac{1}{2}T_R = 5 \cdot 5 t_{RV} = 1 \cdot 36(\tfrac{1}{2}T_I) \qquad (8.14)$$

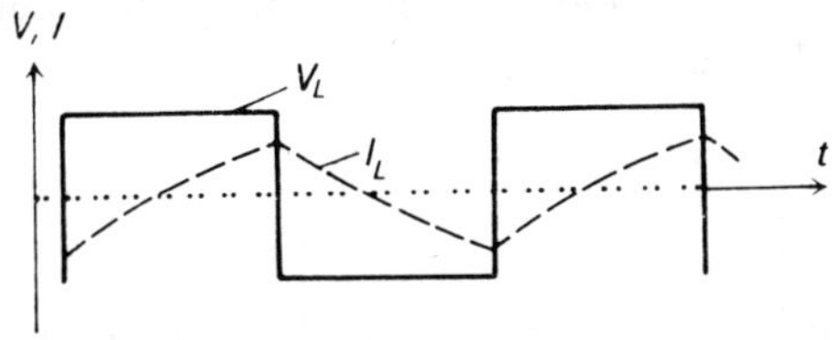

Fig. 146   Load voltage and current for square-wave invertor feeding an inductive load

Note that the load current continues in the same direction as the previous halfcycle for the first part of each new halfcycle of voltage

8.3.2   Inductive load

For resistive load, the load current changes direction as the commutating capacitor begins to recharge in the reverse direction. The overshoot of the capacitor voltage is the result of the energy stored in the d.c. inductor only. The current built up in the inductor during the reverse-voltage time places an extra load on the commutating capacitor, demanding some extra capacitance, but the extra current is generally of the same order as the load current, so that, at the most, the capacitance might be doubled.

Fig. 146 illustrates the lagging load-current waveform for a square-wave invertor. There is a considerable interval at the beginning of each halfcycle during which the load current is flowing in the same direction as in the previous halfcycle. During this interval, load energy is flowing back into the invertor, and, unless other arrangements are made, the commutating capacitor must store this energy, as well as that stored in h, for an acceptable voltage rise. This demands a very substantial increase in the capacitance, bringing also an increased reverse-voltage time. The turnoff process becomes load-related. The increased reverse-voltage time demands that the inductance h is proportionally increased to maintain the current build-up during commutation within bounds. The penalties of operation with an inductive load are thus many, including also poorer voltage regulation, especially with a varying-power-factor load, and a less well-defined output-voltage waveform.

The analysis is clearly more complex since another energy-storage component is present. Three differential equations are required to describe the behaviour, namely

$$
\begin{bmatrix} I'_h \\ V'_C \\ I'_L \end{bmatrix} = \begin{bmatrix} 0 & \dfrac{1}{h} & 0 \\ -\dfrac{1}{4C} & 0 & -\dfrac{1}{4C} \\ 0 & \dfrac{1}{H} & -\dfrac{R}{H} \end{bmatrix} \begin{bmatrix} I_h \\ V_C \\ I_L \end{bmatrix} + \begin{bmatrix} 0 \\ 1 \\ 0 \end{bmatrix} V_s \qquad (8.15)
$$

A full analysis requires the solution of eqn. 8.15 with the matching of initial and final conditions, best accomplished either by building the invertor on an analogue computer which reveals the starting transient and steady-state behaviour, or by an iterative digital-computer program. The reader is also referred elsewhere (Bedford and Hoft, 1964) for a more detailed treatment.

Provided that a square-wave output is acceptable, the problems of the inductive load are best overcome with return-current diodes, described in Section 8.4.

8.4 Return-current diodes

It is clear from Section 8.3 that an inductive load returns energy to the invertor in the first part of each halfcycle, before the load current reverses direction. The conducting thyristor is unable to carry this current, whose direction is suited to the previously conducting thyristor, except via C during the commutation interval. The equivalent of a freewheeling diode is needed to carry this current until it reverses. As a freewheeling diode cannot deal equally with both half-cycles, the solution lies in the use of a fullwave diode rectifier arranged to return load energy to the d.c. supply in the early part of each halfcycle (McMurray and Shattuck, 1961).

The addition of two return-current diodes 1′ and 2′ to a single-phase invertor is shown in Fig. 147. They are positioned to act as a fullwave rectifier feeding load energy into the source of the direct

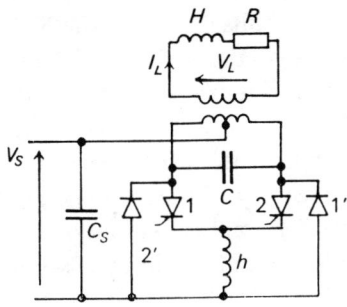

Fig. 147   Return-current diodes 1′ and 2′ added to an invertor to provide a path for load energy to return to the supply during the early part of each halfcycle of voltage

voltage without any intermediate components (e.g. the inductor h) in series. When the d.c. supply cannot accept a reversal of direct current, the capacitor C_s assumes the role of absorbing the load energy returned. (It is less costly to absorb the energy in C_s than in C, because C_s can be a polarised capacitor, and keeping C small permits h also to be kept small.) The return-current diodes as positioned prevent the load voltage from exceeding the supply voltage. The output waveform is therefore 'clipped' at $\pm V_s$, eliminating overshoot and giving a square output voltage waveform. Also, the low value of inductor h reduces the output impedance, improving voltage regulation. The small capacitor C allows the commutation to be thyristor-related, hence fast, reducing losses and giving steep edges to the waveform.

As the analysis of the circuit is complex, the general effects of the diodes are summarised. As indicated in Fig. 146, a rate of change of load current can be expected at the end of the halfcycle, so that not all the supply voltage will be developed across the load, a proportion being dropped across the inductor h. The capacitor C is nevertheless charged to nearly $2V_s$, so that the turnoff mechanism is as described before. At commutation, h has to support a pulse of voltage of amplitude nearly $2V_s$ and duration approximately $2t_{RV}$, adding a buildup current I_{hB} to its initial current. In view of the small capacitance, C is discharged and recharged rapidly, becoming fully recharged to $2V_s$ as one primary terminal falls to zero potential. Any further charging of C is prevented by a return-current diode. If the two half primaries are tightly coupled (e.g. bifilar wound), the load current previously flowing outwards through the right-hand half primary, and, say, through C and 1, can transfer to flow inwards through the left-hand half primary and 2' without any alteration to the transformer m.m.f. Until it reverses, therefore, the load current flows back to the d.c., providing a lossfree recovery of the load energy. While 2' conducts, the left-hand primary terminal is held at zero potential, so that the load voltage is exactly V_s; and the inductor h no longer produces a drop of load voltage.

When 2' begins conduction, the current in h ($= I_L + I_{hB}$) is at its maximum, and cannot change instantaneously. Thus 2', 1 and h continue to carry this current—called a circulating current—which decays slowly since the reverse voltage across h is small, equal to that of two device forward-voltage drops. If the circulating current decays to zero before the load current changes direction (an unlikely assumption, except at very low frequencies), the thyristor 2 will turn off, and will have to be refired when called to carry the load current when it reverses. Invertors with return-current diodes therefore have prolonged gate pulses, lasting at least a quarter of the period, so that gate drive is always present whenever the load current happens to reverse.

In view of the slow decay of circulating current, it is likely that the next commutation will occur while a circulatory current from the previous commutation is still flowing. A cumulative increase of circulating current takes place, until the losses in the two device forward-voltage drops equal the rate of energy injection by commutations. Of greater importance is the fact that the circulating current must be commutated along with the load current, so that the capacitance C must be raised, perhaps by a factor of 2, 3 or more. The device and component current ratings must also be raised to carry the inflated currents without overheating.

The above problems arise in all invertors using return-current diodes. The decrease of circulating current before the next commutation in a lossfree manner is an essential feature of a high-efficiency

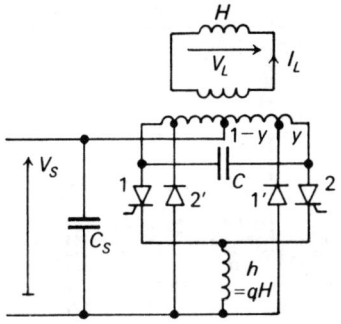

Fig. 148   Return-current diodes 1′ and 2′ connected to tappings a fractional distance y from the ends of the primary towards its centre tap

invertor. Herein lies the design ingenuity behind an efficient invertor circuit.

For the single-phase invertor of Fig. 147, the decrease of circulating current is readily achieved by connecting the return-current diodes to tappings, rather than to the ends of the primary (Fig. 148). The circuit would supply a lagging-power-factor load, but the circuit analysis for this most general case is complex, being divided into intervals over which different sets of differential equations apply. A satisfactory understanding of circuit behaviour can be obtained by considering the purely inductive (zero-power-factor) load, for which the load voltage and current waveforms are more readily drawn since each segment approximates closely to a straight line. Voltage–time areas are now readily calculated, and approximate equations describing the circuit behaviour are derived simply. The solution of these equations is still somewhat complex, and is omitted.

The return-current diodes of Fig. 148 are connected to tappings on each half primary a fractional distance y from the end. For simplicity, the secondary has the same number of turns as a half primary. The inductance h is expressed as a fraction q of the load inductance H. The commutation interval is expressed as a fraction m of the half period $T/2$, and the diode conduction interval a similar fraction n.

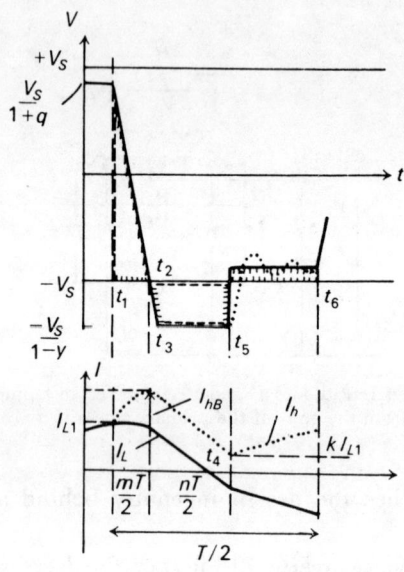

Fig. 149　Voltage and current for Fig. 148 during the halfcycle initiated by firing 1 at t_1

The load-current variation during the commutation interval $mT/2$ is neglected, and the current charging the commutating capacitor during this interval is assumed constant, so that the load voltage changes linearly.

At t_1 in Fig. 149, the end of the previous halfcycle, only thyristor 2 is conducting, so that the currents in the load H and in h are the same, with the same rate of change. The supply voltage is therefore divided between H and h in proportion to their inductances. The load voltage V_{L1} is therefore

$$V_{L1} = V_s/(1+q) \tag{8.16}$$

At the end t_2 of the commutation interval, the load voltage has reversed to $-V_s$. During the commutation interval, the inductor h, carrying an initial current I_{L1}, is subject to the vertically shaded triangular voltage–time area, which increases the current in h by I_{hB}. The increase of I_{hB} during the commutation interval is not influenced by the load inductance, as I_{hB} splits into two halves, one half flowing through one half primary and C, and the other half flowing through the other half primary; thus I_{hB} contributes no m.m.f. to the transformer.

At t_3, a return current diode 2′ begins conduction when the voltage across the section $1-y$ of the half primary has reached V_s. The potential at the end of this half primary is therefore negative, and, since the current in h keeps 1 conducting, a reverse voltage–time area exists across h, shaded horizontally, decreasing the current in h.

At t_4, the current in h has fallen to a value kI_L, and also the load current reverses, ceasing to flow through 2′ and flowing instead through 1. Diode 2′ does not cease conduction until t_5, when the load current has risen (in the reverse direction) to equality with the current in h. From t_3 to t_5, the conduction of diode 2′ has maintained the load voltage constant at

$$V_{L35} = V_s/(1-y) \tag{8.17}$$

From t_5 to t_6, the argument above eqn. 8.16 applies, so that

$$V_{L56} = V_s/(1+q) \tag{8.18}$$

The rate at which the current in h is reduced during the interval t_3 to t_5 is governed by the factor y. A small value of y results in a slow decay, making t_5 extend perhaps to the end of the halfcycle. Still lower values will result in a permanent circulating current, as mentioned earlier. A large value of y can result in t_5 approaching t_4. Still larger values of y will result in the inductor current being collapsed to zero before the reversal of the load current, so that the thyristor will cease and then restart conduction. The load-current flow through 2′ up to t_4 ensures that the load-voltage reduction from $V_s/(1-y)$ to $V_s/(1+q)$ never occurs before t_4. Although this drop in voltage is shown as a step, the capacitance C and circuit inductances make the drop oscillatory, as shown lightly dotted in Fig. 149.

The above description allows the following equations to be written down. Considering first the voltage–time areas across h, we have

$$qHI_{hB} = V_s\left(1+\frac{1}{1+q}\right)\frac{mT}{4} \tag{8.19}$$

and

$$(I_{L1}+I_{hB}-kI_{L1})\,qH = V_s\frac{y}{1-y}\frac{nT}{2} \tag{8.20}$$

and

$$(I_{L1}-kI_{L1})\,qH = V_s\frac{q}{1+q}(1-m-n)\frac{T}{2} \tag{8.21}$$

Considering now the voltage–time areas across H, we get

$$(I_{L1}+kI_{L1})\,H = \frac{V_s}{1-y}\frac{nT}{2} \tag{8.22}$$

and

$$(I_{L1}-kI_{L1})\,H = \frac{V_s}{1+q}(1-m-n)\frac{T}{2} \tag{8.23}$$

Clearly, eqn. 8.22 or 8.23 is redundant, leaving four equations for the four unknowns y, q, n and I_{L1}, taking V_s, H, T, m and k as specified. The parameter $mT/2$ is approximately equal to twice the reverse-voltage time. The parameter k governs the variation of the supply current.

8.5 3-phase centre-tap invertors: with reverse-current diodes

An essential feature of the single-phase centre-tap invertor is the output transformer with its tightly coupled half primaries, allowing the primary current to transfer rapidly from one half to the other, and, in so doing, to transfer to a return-current diode.

Either the 3-phase ($q = 3$) invertor circuit must be similarly provided with bifilar-wound half primaries (Fig. 150), or another d.c. source capable of accepting all the returned load energy, without any intermediate periods of energy output, must be used (Fig. 151). The same technique of connecting the diodes to tappings is used to prevent a circulating current in h. The transformer of Fig. 150 clearly

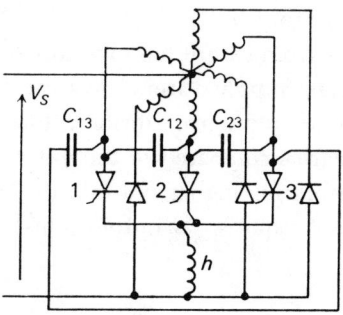

Fig. 150   3-phase centre-tap invertor ($q = 3$) with three additional tightly coupled primaries to provide the return-current diode action

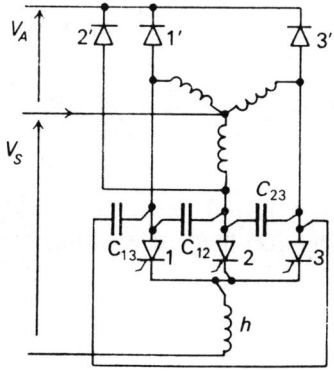

Fig. 151   3-phase centre-tap invertor ($q = 3$) without additional primaries, in which the return-current diodes are connected to another source voltage V_A capable of accepting the returned energy

uses windings wastefully, and, with the addition of three more thyristors and diodes, it would be capable of appreciably more power output. However, if only one thyristor conducts at a time, the form factor of the winding currents will be high. A better conduction pattern would be for three thyristors to conduct at a time, which requires the commutating capacitors to be connected between diametrically opposite thyristors instead of between adjacent thyristors, an arrangement which is identical to three single-phase invertors fed from a common d.c. supply.

Polyphase outputs from centre-tap invertors mostly use a number of single-phase circuits supplied from the same d.c. source and connected to a polyphase output transformer. The increased rating of the centre-tap primaries compared to the output rating favours the use of bridge invertors for higher powers. Bridge circuits have the usual advantage of not requiring an output transformer for the correct

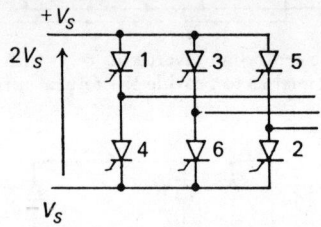

Fig. 152   Basic 3-phase bridge invertor with thyristors numbered
in their conducting sequence

operation of the invertor, and return-current diodes can be added in the form of a bridge rectifier, again without demanding an output transformer.

8.6   Capacitor turnoff: bridge invertors

8.6.1   Self commutation

Although the commutation circuits developed below apply to all bridge invertors, they are introduced with reference to the 3-phase bridge since it conveniently differentiates between two techniques of parallel commutation, each suited to a particular conduction pattern for the invertor thyristors. In Fig. 152, the 3-phase bridge circuit is allocated the usual numbering sequence for its thyristors. If only two thyristors conduct together, the conduction pattern is

$$12$$
$$23$$
$$34$$
$$45$$
$$56$$
$$61$$
$$12 \text{ etc.}$$

from which it is clear that, when 1 is turned off, 3 is fired. Thus commutation occurs between thyristors in the same row or attached to the same d.c. terminal.

If three thyristors conduct together, the conduction pattern is

$$123$$
$$234$$
$$345$$
$$456$$
$$561$$
$$612$$
$$123 \text{ etc.}$$

from which it is clear that the commutation occurs between thyristors in the same column or attached to the same a.c. phase. Two capacitor turnoff techniques are developed from the basic principles to provide the turnoff requirements of these two conduction patterns.

As with centre-tap invertors, so also bridge invertors may have the turnoff process initiated by firing the next load thyristor in the sequence (self commutation) or by firing an auxiliary thyristor (auxiliary commutation). The latter arrangement allows a wider choice of firing sequence since the turnoff and firing processes are now divorced from each other. For instance, the conduction pattern

$$12$$
$$123$$
$$23$$
$$234$$
$$34$$
$$345 \text{ etc.}$$

is readily obtainable using auxiliary commutation.

Returning now to self commutation and the first conduction pattern where firing 3 turns off 1; a capacitor is clearly required between thyristors 1 and 3. When similarly provided for all such pairs of thyristors, the result is as shown in Fig. 153, which is an extension of Fig. 141. The state of charge of C_{13} when 1 and 2 are conducting and prior to firing 3 is governed by the nature of the load, since neither 3 nor 6 is conducting. For a resistive load, the phase b is at a potential midway between the two supply potentials, and hence C_1

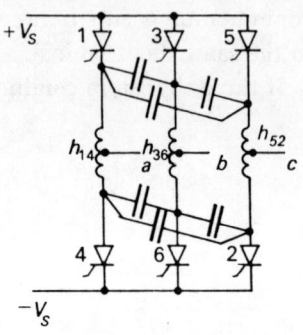

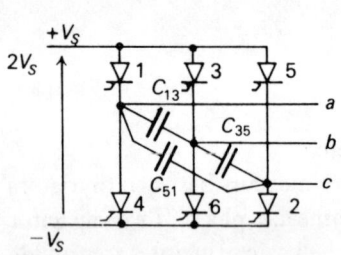

Fig. 153 Primitive capacitor turn-off arrangement which suffers from high di/dt at turnon

Fig. 154 Improved capacitor turn-off arrangement in which the inductors h_{14}, h_{36} and h_{52} reduce the di/dt problem at turnon

Small reactors in series with the capacitors (not shown) are also required to control the di/dt resulting from reverse (carrier-storage) conduction

is charged to V_s (positive or negative). Firing 3 reverse-biases 1 as required. Notice, however, that, when 3 is fired, C_{35} is also charged to V_s (since 2 is conducting). Firing 3 therefore creates a high di/dt through 3, C_{35} and 2, a distinct disadvantage. This is typical of problems which arise in 3-phase capacitor-commutated invertors. One solution is to introduce inductors (Fig. 154).

Consider self commutation for the second conduction pattern, where three thyristors conduct together; it is required that the firing of 4 turns 1 off (Hymphrey, 1968). The capacitor connections of Figs. 153 and 154 are clearly not suited to this function. However, the centre-tapped inductor, with tightly coupled half windings, pro-vides the necessary coupling between the lower and upper thyristor when the potential of its centre tap is restrained by capacitance, as shown in Fig. 155, in which the capacitance can be provided in the two equivalent ways (*a*) and (*b*). Conduction of thyristor 1 ensures that the potential of the phase *a* is near that of the positive d.c. supply terminal $+V_s$, and this potential cannot change instantaneously owing to the capacitors. Firing 4 thus impresses a voltage of amplitude nearly $2V_s$ across the lower half winding, and by transformer action

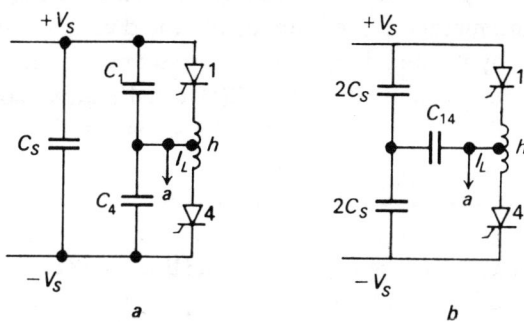

Fig. 155 Basic capacitor turnoff arrangement in which firing 1 turns off 4 and vice versa

Commutating capacitor C_{14} can either be split into C_1 and C_4 and returned to the d.c. terminals as in (*a*), or returned to the midpoint of the d.c. smoothing capacitor C_s as in (*b*)

an equal voltage appears across the upper half winding. A reverse voltage $2V_s$ is thus forced across 1, and the load current I_L previously carried by 1 now flows from the capacitor(s). The inductor current cannot change suddenly, so that the load current which previously was flowing down the upper half winding transfers to, and flows down, the lower half winding, this current also flowing from the capacitor(s). Finally, the inductor itself draws a rising magnetising current on account of the voltage across it. The capacitance must therefore be adequate to carry $2I_L+I_{hB}$ for the duration of the reverse-voltage time. This inductor–capacitor arrangement forms the basis of parallel-capacitor self commutation for many bridge invertors.

8.6.2 Auxiliary commutation

Consider finally the use of an auxiliary thyristor (or thyristors) for turnoff purposes in a 3-phase bridge invertor. Many possibilities exist (Humphrey, 1968; Ohno and Akamatsu 1966); their logical classification is attempted below.

A thyristor in the upper half of a bridge may be turned off by reducing momentarily its anode potential below its cathode potential, or by raising momentarily its cathode potential above its anode potential; these are called, respectively, d.c. and a.c. commutation, since the commutation pulse is applied to the d.c. or a.c. terminal of the thyristor by the auxiliary circuit, at present unspecified.

The auxiliary commutating circuit(s) can be arranged either to reverse-bias all invertor thyristors simultaneously (fully commutated) or to reverse-bias half of the thyristors at any instant (half-commutated) or to reverse-bias only one thyristor at a time (individually commutated).

From the above, six classifications exist (Bradley *et al.* 1964). Some of these inevitably require inductors in the power circuit: all forms of d.c. commutation require inductors, one for a fully d.c.-commutated bridge, two for a half-d.c.-commutated bridge, and six for an individually d.c.-commutated bridge (Figs. 156 *a*, *b* and *c*). A fully a.c.-commutated bridge requires three inductors, to support the voltage which appears at commutation between the positively driven cathodes of the upper-half thyristors, and the negatively driven anodes of the lower-half thyristors (Fig. 157 *a*). A half a.c.-commutated bridge (Fig. 157 *b*) also requires inductors whenever the thyristors in one half of the bridge are reverse-biased; at least one thyristor in the other half must be conducting, thus requiring an inductor to support the potential difference at commutation, as above. A full cycle of operation will clearly require three inductors, but of lower rating than for a fully a.c.-commutated bridge. An individually a.c.-commutated bridge (Fig. 157 *c*) has the unique advantage of requiring no inductors, since it is never required to turn 1 off while 4 is conducting, as 1 and 4 conducting together constitute a d.c. short circuit (McMurray, 1964).

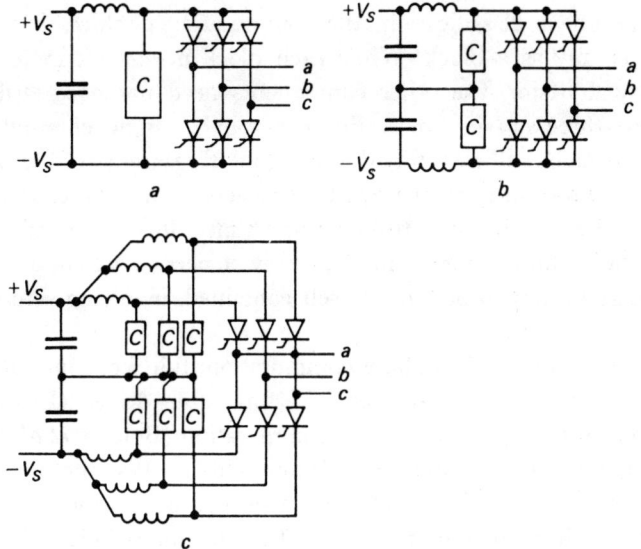

Fig. 156 Auxiliary d.c. commutation circuits C providing (a) fully, (b) half and (c) individually commutated 3-phase bridge invertor

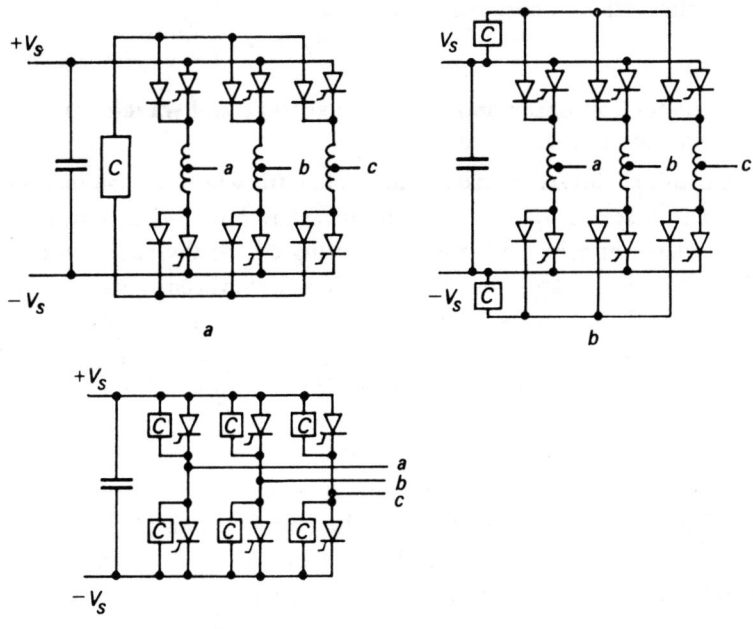

Fig. 157 Auxiliary a.c. commutation circuits C providing (a) fully, (b) half and (c) individually commutated 3-phase bridge invertor

In the above classification, the commutating circuitry has been shown simply as a block. While each block must contain its own auxiliary thyristor, the other components need not necessarily be repeated three or six times. For example, a single commutating capacitor can be sufficient for all six commutating circuits of Fig. 156b, as will be shown in Section 8.9.2. If, for a certain mode, it is desirable to turn off more than one thyristor at a time while using individual commutation for the normal mode, it may be necessary to design each commutation unit to be entirely self contained, involving some cost penalty.

Of the six forms of auxiliary commutation, full d.c. commutation has been used for low-power invertor-motor drives (Li, 1968; Williamson, 1969), while half a.c. commutation (Bradley *et al.* 1964) and individual a.c. commutation (Hammond and Warrington, 1969; Heumann, 1969; Landau, 1969) have been used for medium- and high-power invertor-motor drives. The remaining three forms are not used because of the cost penalty associated with the inductors, which as always bring the associated problems of circulating currents and commutation-energy recovery.

8.7 Conduction paths and patterns: voltage and waveform control in bridge invertors

The bridge invertor has an additional freewheeling conduction path not possessed by the centre-tap invertor. The load current may be circulated at zero load voltage by firing all the thyristors on one side of the bridge and none on the other. Freewheeling paths then exist for both directions of the load current through the thyristors and diodes of one side of the bridge. The three paths possible with bridge invertors are described in Section 8.8 with reference to the single-phase bridge, although they apply equally to polyphase bridges.

A sequence of differing conduction paths is a conduction pattern which generates a corresponding output-voltage waveform; this can be varied by altering the relative conduction durations in each path. Output-voltage control can be obtained by varying the total active period per halfcycle, i.e. the period during which thyristor conduction applies the direct voltage to the load (Jones, 1961; Lawn, 1962; Li, 1968). Output-waveform control can be obtained by varying the relative durations of the active periods within a halfcycle, for example, by modulating (e.g. sinusoidally) the widths of these active periods.

The switching from one path to the next uses the extensions of basic capacitor turnoff developed for bridge invertors in Section 8.6. Certain conduction patterns can be met with self commutation, whereas more complex patterns may be obtained better using auxiliary commutation.

The simplest conduction pattern is a single active period of fixed angular duration per halfcycle, and many invertors use this. As with all fixed conduction patterns, the control of output voltage is not possible. The only way of obtaining a variable output voltage with a fixed conduction pattern is to supply the invertor with a variable direct voltage, which involves a fully rated controlled rectifier or chopper, adding significantly to the cost. The use of a variable direct voltage also introduces a problem when the invertor is required to commutate the full load current with its direct voltage reduced to a low value to start an a.c.-motor load at a low voltage and frequency, for instance. If the commutating capacitor is charged to a voltage proportional to the direct-voltage supply, the commutating ability will fall linearly with the reduction of the supply voltage. An auxiliary constant-voltage supply for capacitor charging must be provided to ensure a fully charged capacitor regardless of direction of the voltage input. The 3-phase invertor described in Sections 8.9.1 and 8.9.2 illustrates the use of an auxiliary voltage to give full commutation ability for a widely varying direct voltage.

8.7.1 For single-phase bridge

Fig. 158a shows a general single-phase bridge square-wave invertor with return-current diodes, omitting the turnoff components. Load current can flow in three distinct paths, depending on which devices are conducting. If devices 3 and 4 are conducting, an active load current flows from the supply, increasing with time. Under no circumstances would a prolonged conduction period include devices 1 and 4 (or 2 and 3), as this would constitute a d.c. short circuit. However, the commutation circuitry can be arranged so that the firing of 1 turns off thyristor 4, thus avoiding the short-circuit possibility (Fig. 155).

Using this technique, firing 1 with a short pulse turns off 4 while the load current continues to flow from right to left. Thyristor 1 cannot carry load current in this direction, so that, after the commutation, 1 ceases to conduct and the diode 4' conducts, forming a freewheeling path with thyristor 3. The load current decays slowly as shown in Fig. 158b, in which the ringed numbers represent the conducting devices and the numbered arrows indicate the thyristor firing instants.

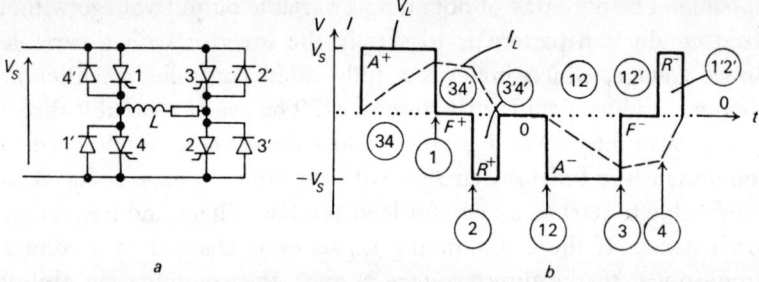

Fig. 158 (a) Basic single-phase bridge invertor with return current diodes feeding an inductive load
(b) Voltage and current waveforms showing active (A), freewheeling (F), return-current (R) and off (O) periods for each current halfcycle
Ringed numbers indicate conducting devices

While the load current still flows from right to left, another conduction path is formed by firing 2, which turns 3 off, leaving 3' and 4' as the only devices capable of carrying the load current in this direction. The conduction of 3' and 4' applies a reverse voltage to the load, hence decreasing the load current rapidly and transferring the load energy to the d.c. source.

When the load current has decreased to zero, no device is conducting, and the load voltage remains zero until 1 and 2 are fired together to initiate the same sequence for the reverse halfcycle.

If these four possibilities are indicated by the letters A (active),

214

F (freewheeling), R (return-current) and O (off), with the superscripts $+$ and $-$ to indicate the load-current halfcycles, the following conduction patterns are readily obtained by the appropriate firing sequence for the thyristors:

(a) Square wave ($A^+R^+A^-R^-$ etc.)

This is the commonest pattern, for which the conduction angles (and hence utilisation) of the thyristors are a maximum. The ratio of output voltage to input voltage is fixed at the maximum, but the harmonic content is not a minimum. The waveforms are shown in Fig. 159.

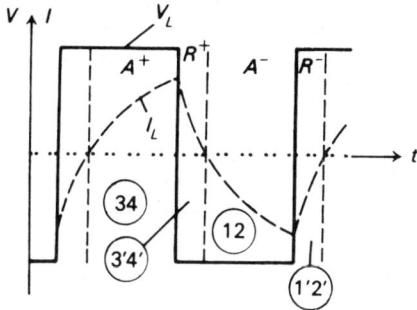

Fig. 159 Output voltage and current for single-phase bridge invertor with inductive load, including active (A) and return-current (R) periods only
Ringed numbers indicate conducting devices

(b) Stepped wave ($A^+F^+R^+A^-F^-R^-$ etc.)

This is a stepped-wave mode for which the output voltage is zero during the freewheeling periods. By making F a variable proportion of each halfcycle, the ratio of output to input voltage is varied (with corresponding variations in the harmonic content: $F = 60°$ eliminates triplen harmonics). This is the commonest mode where voltage control is required. For each direction of load current, there are two possible freewheeling paths. If only one of these is used, the thyristor forming the path carries a larger current than the diagonally opposite one, but there is the control advantage that the firing sequence is identical for each cycle. If the top two thyristors 1 and 3 are chosen for freewheeling duty, the waveforms are as shown in Fig. 160, in

215

which the phase relationship between the firing pulses for 1 and 4, and 2 and 3, is fixed, and voltage variation is achieved by shifting pulse 4 with respect to pulse 1, and pulse 2 with respect to pulse 3. If the left-hand thyristors 1 and 4 are chosen for freewheeling duty, the waveforms (Fig. 161) give an identical output waveform. The pulses to 1 and 4 are a fixed 180° apart, as are the pulses to 2 and 3,

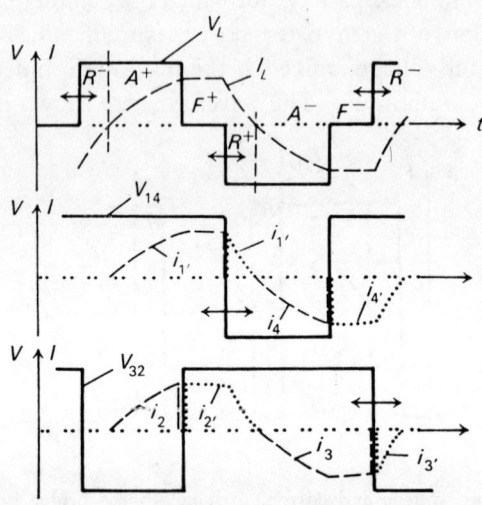

Fig. 160 Output voltage and current for single-phase bridge invertor with inductive load, including active (A), freewheeling (F) and return-current (R) periods

Output-voltage variation is obtained by varying the termination of the F periods as indicated by arrows

and voltage variation is now achieved by phase variation ϕ between the pulses to 1 and 4, and 2 and 3.

If the freewheeling action is distributed equally between all paths, the firing sequence occupies two cycles, and the waveforms follow Figs. 160 and 161 alternately, as shown in Fig. 162.

The phase-shifting between the pulses to the left-hand thyristors 1 and 4 and right-hand thyristors 2 and 3 is the most common technique for single-phase bridge invertors, and it can be applied also where there are multiple commutations per halfcycle, as described below. It is not, however, possible for 3-phase bridge invertors, as

216

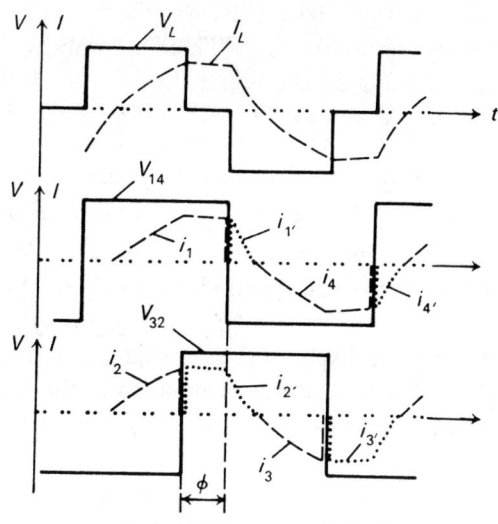

Fig. 161   Output voltage and current for single-phase bridge invertor
with inductive load, with A, F and R periods

Output-voltage variation is obtained by variation of ψ,
the phase shift between the square waves V_{14} and V_{32}

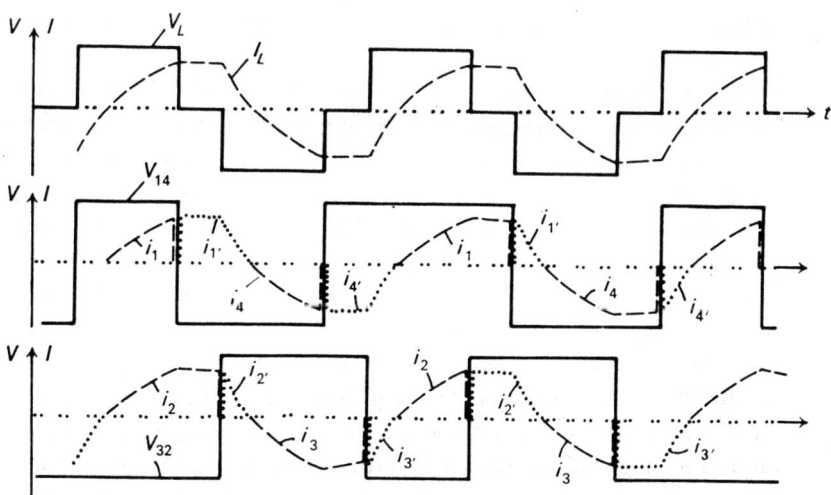

Fig. 162   Conduction pattern lasting two cycles which equalises the current
ratings of all the devices in the invertors of Figs. 158–161

the firing pulses to the thyristors feeding the three phases must always be separated by a constant 120°. There are only two available free-wheeling paths, using all the upper thyristors and diodes or all the lower thyristors and diodes. This is developed in Section 8.7.2.

(c) Modulated wave

$$(A^+F^+A^+...F^+R^+F^+R^+A^-F^-A^-...F^-R^-F^-R^-A^+...etc.)$$

This mode has multiple commutations per halfcycle. By modulating the width of successive active periods sinusoidally, the low harmonics present in a square wave can be much reduced, leaving only the fundamental which shapes the load current, and high harmonics which contribute a small ripple component to the load current.

The same choice of paths exists for each freewheeling period as above. If each voltage halfcycle is restricted to containing unidirectional supply voltage periods or freewheeling periods only, i.e. there are no periods of negative load voltage during the positive-voltage halfcycle, a conduction pattern developed from Fig. 160 is used. The two thyristors feeding one load terminal are given a suitably chosen, fixed conduction pattern; the conduction pattern of the two thyristors feeding the other load terminal is varied, as shown by the arrows of Fig. 163, to give the output waveform and waveform variation shown.

By varying the conduction pattern of both pairs of thyristors 1 and 4 and 2 and 3 in the same manner but with a phase shift between the modulating signals, voltage control and modulation can be obtained together. Fig. 164 shows the result of using two triangular modulating waveforms phase-shifted by 67°; one modulating waveform determines the instant of firing for thyristor 1 and the other controls thyristor 3 similarly. Thyristors 4 and 2 are fired together at regular intervals, shown by the synchronisation of the negative-going edges of the V_{14} and V_{32} waveforms. The voltage can be controlled by varying either the amplitude or the phase displacement of the modulating signals, which are usually triangular. The firing instants are generated here by comparing the triangular modulating signal with a sawtooth waveform, whose steep edge determines the synchronised firing of thyristors 4 and 2.

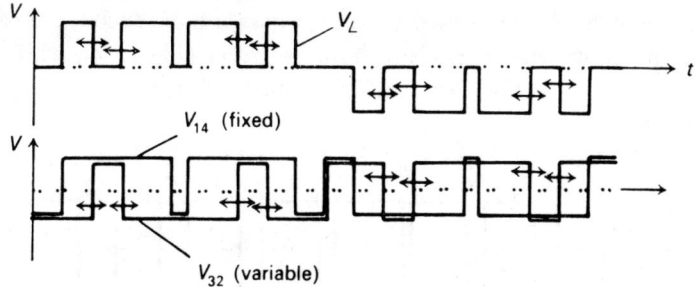

Fig. 163 Voltage and waveform control produced by a form of modulation, in which the timing of the vertical edges marked with arrows is adjustable

The waveform of V_{14} at one load terminal is fixed while that at the other end, V_{32}, is varied as shown to produce the modulation

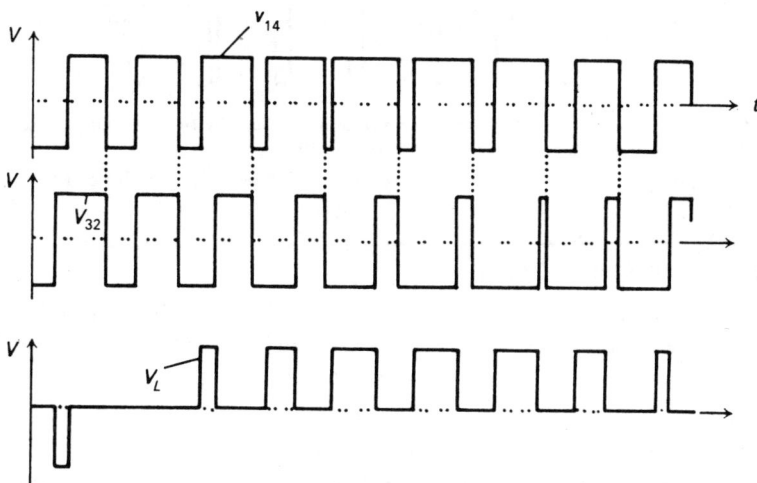

Fig. 164 Voltage and waveform control produced by the modulation of firing instants of thyristors 1 and 3 by two triangular waveforms phase-shifted by 67°

Thyristors 4 and 2 are fired together at regular intervals, thus synchronising the negative-going edges of V_{14} and V_{32}

If instead the firing of thyristors 4 and 3 is synchronised, the negative-going edge of V_{14} coincides with the positive-going edge of V_{32}. A modulated output voltage waveform is still produced, but each halfcycle has varying periods of both polarities (Fig. 165), which is less desirable from the viewpoint of harmonic distortion.

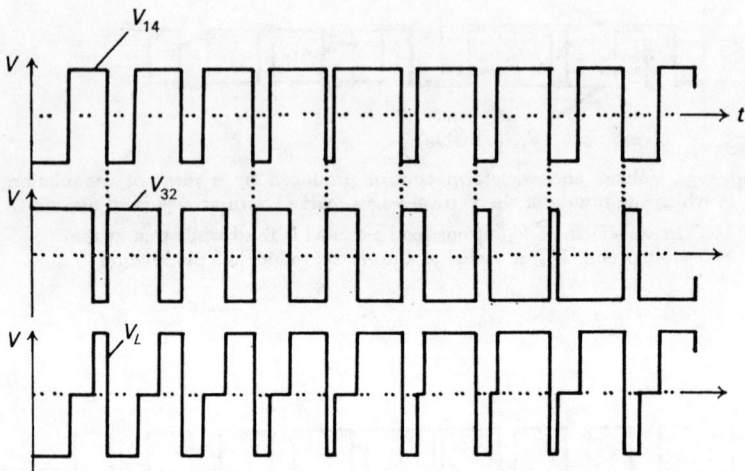

Fig. 165   Modulated output waveform obtained when the negative-going edge of V_{14} is synchronised with the positive-going edge of V_{32}

Each output halfcycle is now composed of varying periods of both polarities

8.7.2   For 3-phase bridge

The 3-phase bridge has more scope for active conduction patterns, since the load may be energised by the conduction of two thyristors or three simultaneously. These two patterns have already been introduced in Section 8.6.1, where the basic turnoff mechanisms appropriate to each were developed. The freewheeling loops, as for the single-phase bridge, occur in the upper or lower halfbridge, and may freewheel all three phase currents simultaneously, or may be chosen to freewheel a selected pair only.

Whereas the single-phase bridge offered two forms of voltage control by varying the total active period per halfcycle, the second form (employing a phase shift between the firing signals for the thyristors feeding one load terminal compared with those for the other thyristors) is not allowable for 3-phase invertors, as the corresponding phase shift must be permanently fixed at 120°. This is an important restriction and, where the best possible waveshape control is required, may lead to the use of three single-phase bridges in preference.

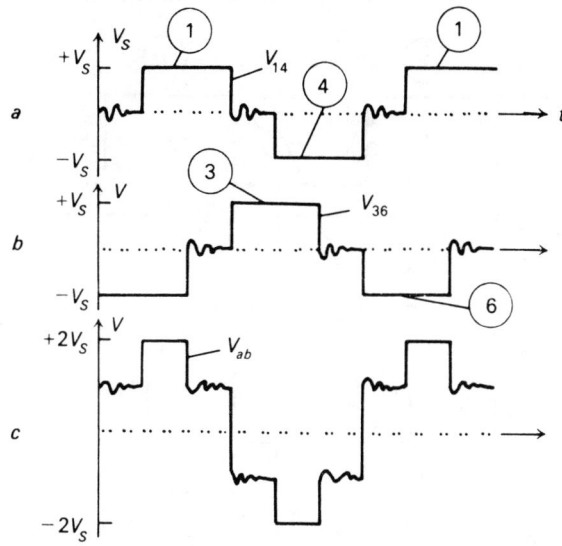

Fig. 166   Phase-potential waveforms (*a*) and (*b*) for 3-phase bridge invertor in which two thyristors only conduct together; the corresponding line–line voltage V_{ab} is shown at (*c*)

The oscillatory parts of the waveform arise where a phase is energised by neither thyristor

(*a*) *Square or stepped wave*

The 2-conduction and 3-conduction patterns give rise to two possible output waveforms, neither of which allows voltage control or current waveshaping within the invertor.

With two thyristors conducting together, the conduction angle is 120°, and the phase potential waveform with respect to the positive and negative supply potentials is shown in Fig. 166. Of the six invertor periods in one cycle, there are two when neither thyristor supplying a phase is conducting, giving an undefined phase potential.

The line–line voltage waveform (Fig. 166c) also exhibits periods of undefined voltage at approximately half the peak output voltage. For this reason, and because the 120° conduction angle gives slightly poorer thyristor utilisation, a conduction angle of 180° is preferred.

Three thyristors conducting together give a conduction angle of 180°, and the well defined phase potential waveform shown in Fig. 167

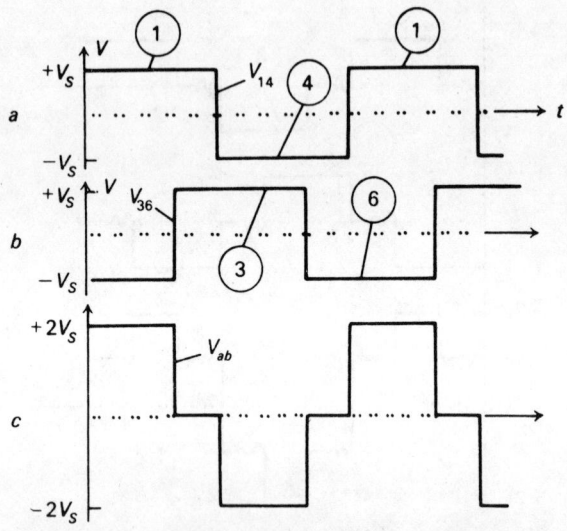

Fig. 167   Phase-potential waveforms (a) and (b) for 3-phase bridge invertor in which three thyristors conduct together; the corresponding line–line voltage V_{ab} is shown at (c)

with its corresponding line–line voltage waveform. The harmonic content of this and of the earlier line–line waveform is identical, neither containing triplen harmonics and each having 1/5th of fifth, 1/7th of seventh, 1/11th of eleventh, 1/13th of thirteenth etc. Both conduction patterns yield an output voltage the mean value of which is two-thirds of the direct supply voltage with a fundamental component of amplitude $(2\sqrt{3}/\pi) \times$ (direct supply voltage). Fig. 168 gives the fundamental and higher harmonic components for output waveforms with active periods of varying angular durations.

(b) *Stepped wave*

Where voltage control within the invertor is required (to eliminate the controlled rectifier or chopper for direct-voltage control), the simplest approach is to fire the thyristors for the variable latter part of each invertor period, making the latter part 'active'. The early part

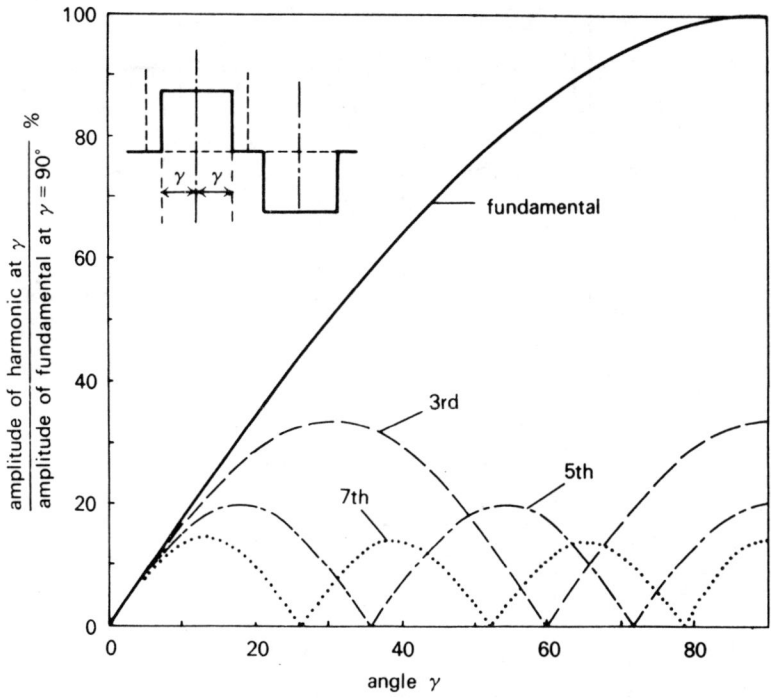

Fig. 168 Amplitude of harmonics present in a stepped wave, expressed as percentage of amplitude of fundamental present in square wave ($\gamma = 90°$)

of each invertor period can either have no thyristors conducting, in which case the appropriate return-current diodes conduct and the load currents collapse rapidly, or all the thyristors on one side may be fired, establishing the freewheeling paths for the load current and giving a slower decay. These alternatives are shown in Figs. 169 and 170 for the case where the thyristors conduct for 75% of each

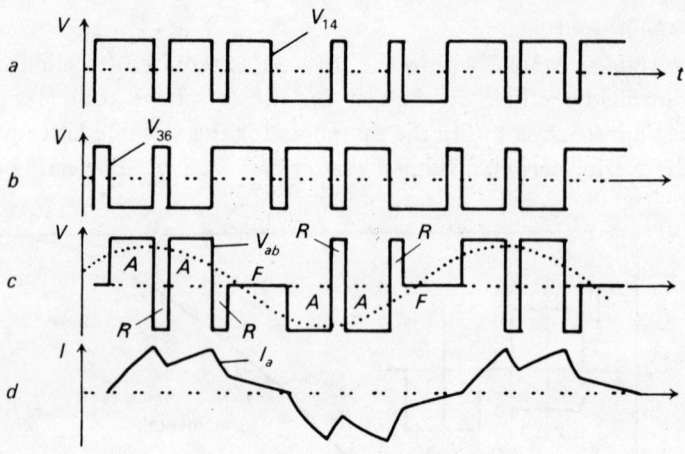

Fig. 169 Potential waveforms for (a) V_{14} and (b) V_{36} in a 3-phase bridge invertor in which the first 25 % of each invertor period is a return-current period (R) and the latter 75 % in active (A); line–line voltage V_{ab} (c) and line-current I_a waveforms d) are also shown

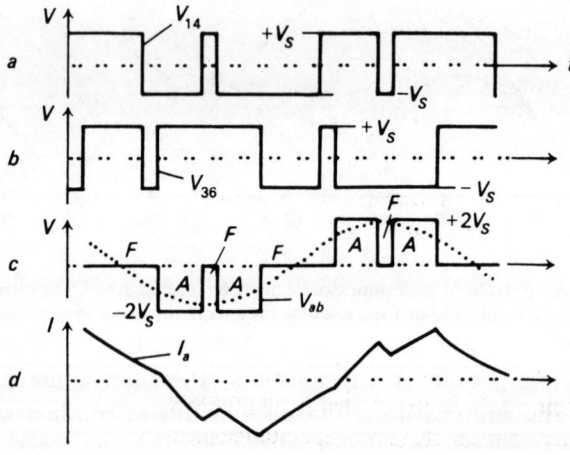

Fig. 170 Potential waveforms for (a) V_{14} and (b) V_{36} in a 3-phase bridge invertor in which the first 25 % of each invertor period is a freewheeling period (F) and the latter 75 % is active (A); line–line voltage waveform (c) and line-current waveform (d) are also s hown

224

invertor period; it is clear that the latter has a lower harmonic content in the output voltage, and consequently has a smoother load current (Mokrytzki, 1967).

(c) Modulated wave

When a.c. motors are fed at low frequency from an invertor, the square or stepped waveform is insufficiently smoothed by the motor inductance to give smooth rotation. The current waveform has noticeable steps after each commutation, so that the flux wave rotates with

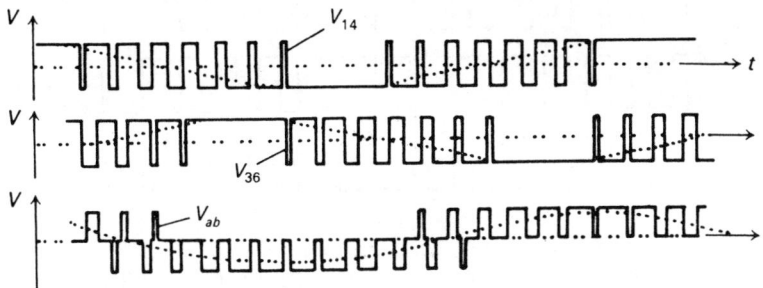

Fig. 171 Modulated output voltage V_{ab} obtained using trapezoidal modulation by mark/space-ratio variation of the phase potential waveforms V_{14} and V_{36} which are phase-displaced by 120° to each other, as required for the 3-phase-bridge circuit

uneven angular velocity, giving 'jerky' shaft rotation. Multiple commutations with sinusoidally modulated 'active' periods give much smoother rotation.

It is generally required to provide voltage control as well as sinusoidal modulation, which restricts the conduction pattern considerably. One technique which does not involve an excess of control complication is the generation of a trapezoidally modulated phase potential waveform (Fig. 171). The sloping region is obtained by variation of the mark/space ratio. Note that the phase-b potential waveform, V_{36}, is shifted by 120° from that of phase a, V_{14}, as required, while the line–line voltage waveform obtained by subtraction exhibits the pseudosinusoidal modulation required. The waveforms are shown for maximum output voltage; for reduced voltage, the mark/space-ratio control does not span the extreme range shown in Fig. 171

during the modulated periods of V_{14} and V_{36}. Instead, the modulated periods begin and end with a much wider pulse of opposite polarity, and the intermediate change is consequently less. Minimum voltage occurs where the mark/space ratio is constant at $1:1$ throughout the 'modulated' period. For this extreme condition, the effect of modulation is clearly lost since there is no change in mark/space ratio, and the mean output voltage is reduced to approximately half that corresponding to Fig. 171.

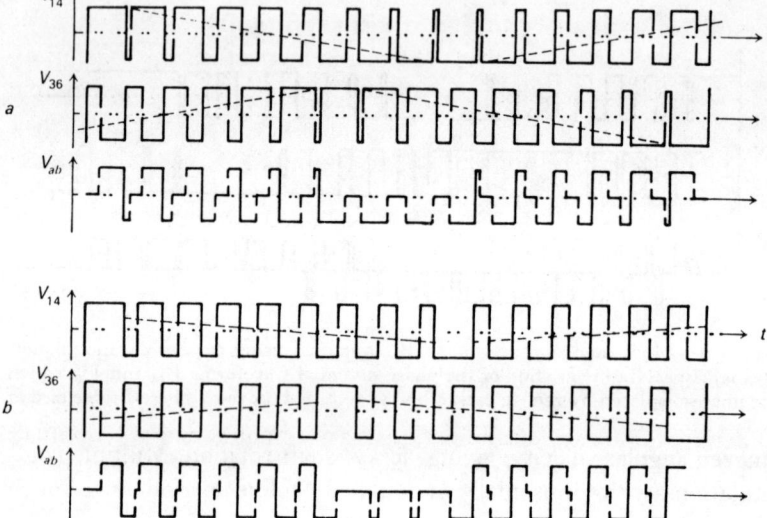

Fig. 172 Modulated output voltage V_{ab} for (a) full voltage and (b) for a lower mean output voltage of about 25 %, when triangular modulation of mark/space ratio is used. Note how the amplitude of the fundamental is reduced and the higher harmonic content increased for the lower voltage output

For a wider range of voltage control, the duration of the flat tops of the trapezoidal modulating waveform must be shortened, yielding in the limit a triangular modulating waveform. With this waveform, the output voltage can be reduced nearly to zero by reducing the range of modulation. As before, however, the output voltage waveform loses much of the beneficial shaping as the range of mark/space-ratio variation is reduced. Fig. 172 illustrates an approximation to

226

triangular modulation, at maximum output voltage (Fig. 172*a*), at which the harmonic contact is noticeably greater than in Fig. 171, and at low voltage (Fig. 172*b*), at which the effects of modulation are almost lost.

8.8 Single-phase bridge invertors

The single-phase invertor is used mainly for static power supplies, usually battery-supported, where the range of voltage control is small. It supplies a constant output voltage for a range of input voltages from maximum battery voltage to about two-thirds of this value. For powers above 5 kW, the bridge circuit is better than the centre tap. As such invertors power high-priority monitoring and electronic equipment, a sine-wave output is required, typically with less than 5 % total harmonic distortion, 1 % voltage and $\frac{1}{2}$ % frequency regulation. The sine-wave output may be obtained by using the bridge equivalent of the centre-tap invertor with load-related parallel-capacitor commutation, or with series commutation (Section 7.5.1), or by using a square-wave invertor and filter. In view of the similarity between the operation of the centre-tap and bridge invertors with parallel commutation (each has the same equivalent circuit), no further discussion is given. A practical square-wave invertor using self commutation is now examined; auxiliary commutation (Section 8.9) is applied to two 3-phase invertors.

8.8.1 Practical single-phase bridge square-wave invertor

Having surveyed the possible conduction paths for an invertor, the detailed operation of a practical circuit capable of operating in any of the above modes is considered. The circuit, published elsewhere (Brisby, 1965), has been chosen because it illustrates the commutation process in its basic form, and introduces the problem of circulating currents together with a practical solution. The circuit is reproduced in Fig. 173.

The invertor thyristors are 1, 2, 3 and 4 with their return-current diodes 1′, 2′, 3′ and 4′. The commutating inductors h_{14} and h_{32} have tightly coupled halfwindings, and, together with C_1, C_2, C_3 and C_4, enable the firing of one thyristor to turn off the other thyristor connected to the same inductor (Fig. 155).

Assume that 1 and 2 are conducting: V_{14} is at the potential $+V_s$ and V_{32} at $-V_s$, neglecting the small voltage drops across h. When 3 is fired, a voltage $2V_s$ is forced across the upper half winding of h_{32}, which reproduces an equal voltage across its lower half-winding, reverse-biasing 2 with a voltage $2V_s$. The load current which flowed previously into 2 now flows into the combined capacitor C_{32}; also the

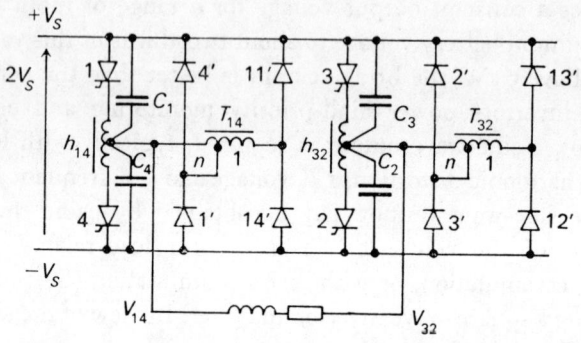

Fig. 173 Practical single-phase bridge square-wave invertor using self commutation, in which the commutation energy is recovered via the transformers T_{14} and T_{32} with their associated diodes

load current flowing in the lower half of h_{32} now flows in the upper half of h_{32} and into C_{32}, doubling the capacitor current. In addition, a rising magnetising current flows through 3 and the upper half of h_{32}, this current also flowing into the combined capacitor C_{32}. The potential of V_{32} rises as these three currents charge C_{32} positively until the return-current diode 2′ prevents any further rise of voltage by accepting the capacitor current itself.

If the return-current diodes were connected directly to the centre taps of the inductors h, the capacitor voltage would be clipped at $+V_s$ by 2′, but the magnetising current in h_{32}, now at its peak, would have a freewheeling loop through 2′ and 3, allowing it to decay only slowly. Large circulating currents would impose a severe additional load on the commutating circuits, demanding components of increased size and cost.

As it is, the potential V_{32} can rise above V_s by a voltage $2nV_s$ where T_{32} has $n+1$ turns with n turns between $2'$ and h_{32}. A reverse voltage thus appears across h_{32} to decay its magnetising current, while all the currents previously flowing in C_{32} now flow through $2'$ and n turns of T_{32}. T_{32} demands an opposing m.m.f. which flows through $12'$ and $2'$, thus fixing the voltage across this section of winding at $2V_s$, and by transformer action fixing the voltage across the section n at $2nV_s$.

Thus the energy injected into h_{32} at commutation is transferred partly to the load as a higher load voltage, and partly returned to the supply through the diodes $12'$ and $13'$. Assuming the recovery of energy in h is complete before the load current reverses, 3 will cease conduction, leaving the load a freewheeling path which includes here a reverse voltage $2nV_s$, through 1, T_{32} and $2'$. The freewheeling mode has been established.

If now 4 were fired to turn off 1, and the energy in h_{14} is fully recovered before the load current reverses, a return-current path ($1'$, T_{14}, load, T_{32}, $2'$ and supply) exists which reduces the load current, returning the load energy to the d.c. supply.

The circuit arrangement of Fig. 173, using only four thyristors, allows the various modes of operation to be used, but with some restrictions, based on the time to recover the energy injected into h at each commutation. If a close approximation to a square wave is required, the factor n must be small, which increases the energy-recovery time, and hence the time required for the circuit to transfer from the active mode to the freewheeling mode, for instance. This, and the need for the load-carrying inductors h and the return-current transformers T, has led to the development of faster circuits which avoid these bulky components, but which, using auxiliary commutation, require additional thyristors. The cost penalty of the extra thyristors is becoming progressively less significant.

To avoid duplication, the single-phase bridge auxiliary-commutated invertor is not described, as the 3-phase version is described at the end of Section 8.9.

8.9 3-phase bridge invertors

The possible conduction paths and commutation techniques have already been described in Sections 8.6 and 8.7. A self-commutated 3-phase bridge invertor is practicable, but it follows exactly the description of the single-phase version, where the self-commutation circuitry is appropriate to the preferred 3-thyristor conduction pattern for the 3-phase bridge. This Section is consequently devoted to auxiliary commutation, with examples of three circuits, two for variable direct voltage and the third for variable output voltage from a fixed direct voltage by multiple commutations and waveform modulation. This choice is appropriate to a common application for 3-phase bridge invertors in which the invertor supplies directly the windings of a 3-phase variable-speed motor (Bradley *et al.* 1964; Heumann, 1969; Landau, 1969); operation at low speed requires reduced voltage, and occasionally the waveform-modulation feature to obtain smooth rotation.

A 3-phase load having electrically isolated phases can be simply (but rarely economically) supplied from three single-phase invertors, their firing signals displaced by 120°, fed from a common d.c. supply. Apart from the influence of mutual inductance between load phases on phase current, each invertor behaves as described earlier. The arrangement requires 12 thyristors for three single-phase bridges, compared with six for the more economical 3-phase bridge.

8.9.1 Variable-direct-voltage auxiliary d.c.-commutated invertor

First, an invertor using half d.c. commutation and a variable direct voltage is examined. The objective is to use the charging of the commutating capacitor following the commutation of one-half of the bridge to leave the capacitor at the correct voltage and polarity for the subsequent commutation of the other half of the bridge; further, the initial reverse voltage at turnoff should be independent of the direct-voltage supply. Immediately after a commutation, the invertor circuit voltages are fixed by the return-current diodes which, for d.c. commutation, are returned directly to the d.c. supply (Fig. 174).

At the end of the commutation which turned off 1, 3 and 5, the current in h_{11} (the load current just before commutation plus the rise ·in magnetising current generated by the commutation itself) has charged C until the anode of 11 has risen to the potential $+V_s$, at which 8′ takes over the current in h_{11} so that 11 turns off. Owing to the presence of the auxiliary voltage (shown as a battery, but in practice a transformer rectifier), the right-hand plate of C is at a potential V_s+V_A, and hence the anode of 12 is at a potential V_s+2V_A. Thus, when 12 is fired to turn off thyristors 4, 6 and 2, the cathodes of 4, 6 and 2 are forced to a potential V_s+2V_A, while their anodes are

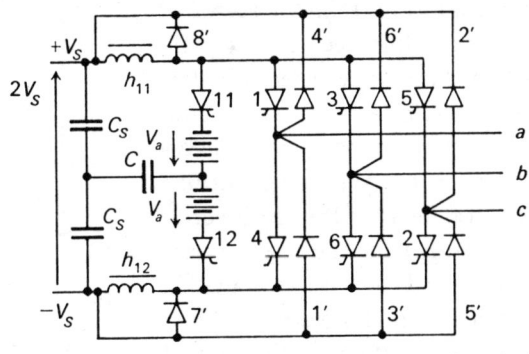

Fig. 174 Practical 3-phase bridge invertor using the principle
of half d.c. commutation

Auxiliary voltage V_a allows the invertor to operate at
rated current for all values of the supply voltage $2V_s$

clamped at $+V_s$ by the return current diodes 4′, 6′ and 2′. The initial reverse voltage across these thyristors is thus $2V_A$, which is independent of the supply direct voltage $2V_s$ as required.

The major difficulty with invertors using d.c. commutation and thus requiring inductor(s) between the d.c. supply and the commutating circuits, is the inability of the inductor, when spanned by a freewheeling circuit, to discharge the energy which is repeatedly injected into it at each commutation. For energy recovery, a reverse voltage must be allowed to occur across the inductor to give a reverse-voltage–time area before the next commutation greater than the voltage–time area which injected the energy. A secondary winding for h_{11} and h_{12} has been suggested which feeds back to the d.c. supply through a diode, with the removal of 8' and 7'.

For a reverse voltage across h_{11} of 10%, the turns ratio must be 10 (secondary) : 1 (primary), so that, when at commutation the primary supports over twice the supply voltage, a voltage of over 20 times the supply voltage is induced in the secondary, which must be withstood by the diode. Some benefit can be had by returning the secondaries to half the primary voltage, making use of the two series portions of C_s (Fig. 175).

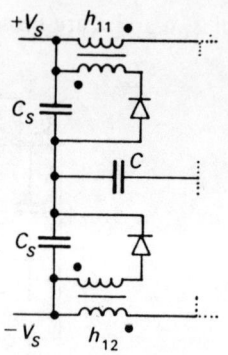

Referring again to Fig. 174, it is seen that 8' is not the only possible freewheeling path, as 1 and 4' etc. can also freewheel the current in h_{11} if 1 is refired immediately after a commutation. Correct energy recovery thus depends on a delay between turning off 1, 3 and 5 at the end of the 123 invertor period, and refiring 3 for the 234 invertor period which follows. Whereas this limitation is not too important in an invertor employing only one commutation at the end of each invertor period, it seriously limits the use of d.c. commutation for multiple commutations within an invertor period.

Fig. 175 Suggested solution to the problem of extracting the energy injected into the d.c. reactors h_{11} and h_{12} at each commutation

8.9.2 Variable-direct-voltage auxiliary a.c.-commutated invertor

Secondly, an invertor using half a.c. commutation with a variable direct voltage is examined. Again the objective is to use the recharging following one commutation to leave the capacitor at the correct voltage for the next. A practical circuit is shown in Fig. 176. During the commutation of 1, 3 and 5, the load currents were transferred to 1', 3' and 5', then to 11 and C_{11}, but when the upper plate of C_{11} has fallen to, and is clamped at, the potential $-V_s$ by 7', the auxiliary

232

voltage ensures that the lower plate of C_{12} is at a potential $-V_s-V_A$ (i.e. C_{12} is charged to V_A), so that, at the next commutation, firing 2 will force a reverse voltage V_A across 4, 6 and 2 as required. This reverse voltage is independent of the supply voltage $2V_s$.

As before, the commutation process charges a commutation choke with energy, which, for efficient operation, must be recovered. The full mechanism is explained in detail elsewhere. The largest injection of energy occurs in h_{52} at the end of the 123 invertor period. At this

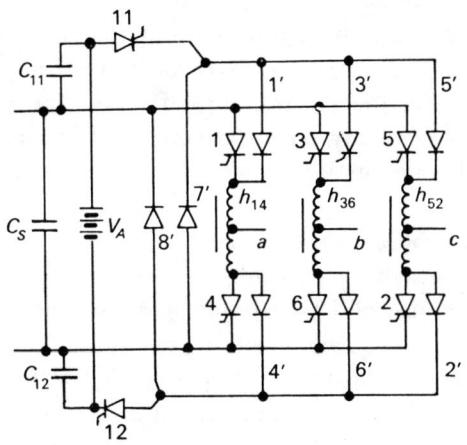

Fig. 176 Practical 3-phase bridge invertor using half a.c. commutation

Auxiliary voltage V_A allows the invertor to operate at rated current for all values of supply voltage $2V_s$. The commutation energy injected into h_{14}, h_{36} and h_{52} is readily recoverable

commutation, the upper end of h_{52} is raised to a potential $+V_s+V_A$ while its lower end is held at $-V_s$ by the conduction of 2. The large pulse of voltage raises the current in h_{52} substantially, and this current circulates throughout the next (2, 3 and 4) invertor period, flowing freely through 2, 7' and 5'. At the next commutation, however, 4, 6 and 2 are turned off, after which the load current which flowed previously through 2 to $-V_s$ now flows through 2' and 8' to $+V_s$. The reversal of phase potential causes the load current to decay. Provided that the firing of 5 in the next invertor period (3, 4 and 5) is delayed, the current circulating in h_{52} in the previous invertor period (2, 3

233

and 4) has no alternative but to flow through 2′ and 8′ to $+V_s$, and from $-V_s$ through 7′ and 5′. A large reverse voltage $2V_s$ thus appears across h_{52}, reducing the current in it rapidly and returning the commutation energy to the d.c. supply. Only a small firing delay is required to recover the commutation energy, so that 5 can be fired before the reversal of load current and the delay thus has no influence on the inverter output waveforms. The much faster energy recovery of this circuit represents a major advance over half d.c. commutation.

Each of the circuits above has its commutation based on forcing a large reverse voltage across the thyristors to be turned off, with the object of obtaining the shortest turnoff time. The return-current diodes have consequently not been connected in parallel with the thyristors, and commutating inductors have been included to allow the reverse voltage to be applied to the thyristors in spite of the return-current diodes. A sudden rise of reverse voltage across the thyristors of one half of the bridge, when a.c.-commutated, introduces a sudden rise of forward voltage across the nonconducting thyristors in the other half. Small capacitors (with some series resistance to limit the inrush current) in parallel with the thyristors, in conjunction with the commutating inductors, control this dV/dt adequately.

8.9.3 Fixed-direct-voltage variable-output-voltage modulated invertor using individual a.c. commutation

The treatment of 3-phase bridge invertors is concluded with an example using individual commutation and having the return current diodes directly in parallel with the thyristors (Hammond and Warrington, 1969; Heumann, 1969; Landau, 1969). Although the turnoff time is increased slightly by restricting the reverse voltage at turnoff to the forward-voltage drop of a diode, the above dV/dt problem is avoided. Most important, the inductors are no longer required. The circuit is suitable for star or delta loads, and is capable of multiple commutations per invertor period providing a modulated waveshape. Originated by McMurray (McMurray, 1964), it has been widely used for variable-frequency invertor-motor drives, where both voltage control and a modulated waveshape are required for smooth motor rotation at low frequencies (speeds) and low voltages. At higher speeds, the modulation feature may be omitted, thus reducing the commutation rate and hence the commutation losses.

The circuit is shown in Fig. 177, in which the load thyristors are 1–6, the return current diodes are 1'–6', and the auxiliary thyristors are 11–16. Commutation capacitors C_{14}, C_{36} and C_{52} are provided with resonant inductors l_{11}–l_{16}.

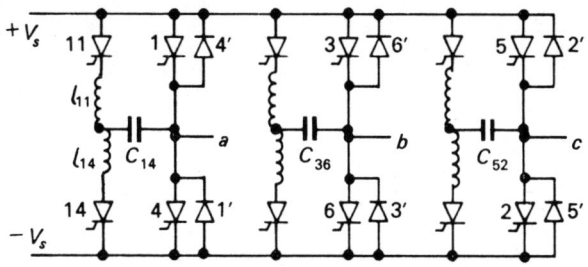

Fig. 177 Practical 3-phase bridge invertor using individual a.c. commutation, and operating from a fixed-voltage d.c. source

Output-voltage variation and waveform modulation are both obtainable with this circuit

Whether stepped-wave or modulated-wave operation is used, the commutation process always involves turning off one thyristor by firing the auxiliary thyristor connected to the same phase and d.c. terminal: 11 turns off 1 etc. The initial voltage on the capacitor is approximately $4V_s$ (with polarity − +); it is shown how, at the end of the process, the capacitor is left with this voltage and the correct polarity for the next commutation. Thyristor 1 is conducting the load current; to turn 1 off, 11 is fired, which initiates a resonant current through 11, l_{11}, C_{14} and through 1 in opposition to the load current. The peak resonant current I_{Rp} is chosen so that

$$I_L = k_i I_{Rp} \quad \text{where generally} \quad \tfrac{1}{2} \leqslant k_i \leqslant \frac{1}{\sqrt{2}} \qquad (8.24)$$

Taking, for example, $I_L = \tfrac{1}{2}I_{Rp}$, 30° of resonant oscillation is required for the resonant current to become equal and opposite to the load current in 1, and hence to reduce I_1 to zero. For the next 120°, the excess of resonant current (over the load current) flows through diode 4′. At 90°, the peak of I_R (Fig. 178*b*), the capacitor is discharged and thereafter it charges with reverse polarity $(+ -)$. When the resonant current has fallen to equality with the load current at 150°, the current in 4′ has fallen to zero, and any further reduction of resonant current means that 1′ must begin conducting, to make up the

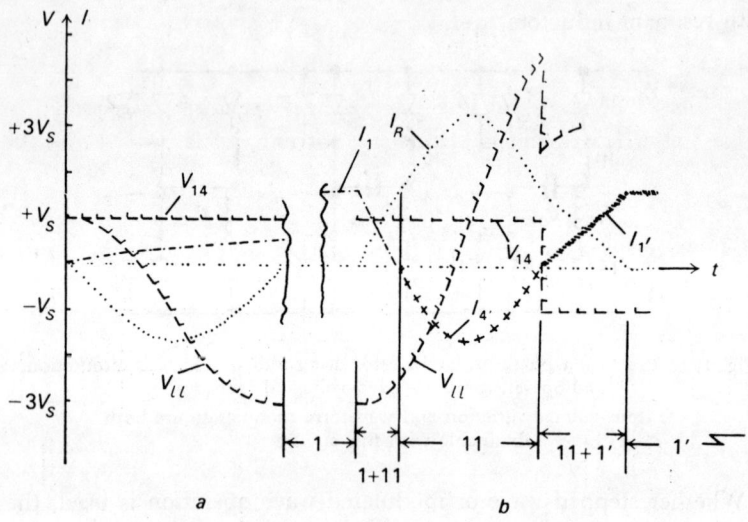

Fig. 178 Resonant turnoff waveforms for Fig. 177

    a Initial capacitor charging process
    b Voltage and current waveforms during normal turnoff process

difference between the load current (assumed constant) and the reduced resonant current. As 1′ begins conduction, 4′ ceases to conduct, and the phase potential V_{14} switches from $+ V_s$ to $- V_s$, altering the configuration of the resonant circuit, and reducing the voltage across l_{11} by $2V_s$. Before the change of phase potential, the rising potential on the left-hand plate of C_{14} was causing an increasingly rapid decay of resonant current. After the change of phase potential, the resonant current still decays resonantly, but at a slower rate as shown. When the resonant current has decreased to zero, the load current flows entirely through $1'$, and the left-hand plate potential of C_{14} is $+3V_s$, giving a capacitor voltage $4V_s$ ($+$ $-$). For the chosen ratio,

$$k_i = I_L/I_{Rp} = \tfrac{1}{2} \tag{8.25}$$

and, defining k_V as

$$k_V = V_{Cp}/2V_s \simeq 2 \tag{8.26}$$

l and C are calculated from the following equations:

$$t_{RV} = \pi\sqrt{(lC)}\,(1 - \tfrac{1}{90}\sin^{-1}k_i) \tag{8.27}$$

$$C(2k_V V_s)^2 = l\left(\frac{I_L}{k_i}\right)^2 \tag{8.28}$$

The phase potential and line–line voltage waveforms are shown in Fig. 179 with the addition of the load current, 30° lagging. From the

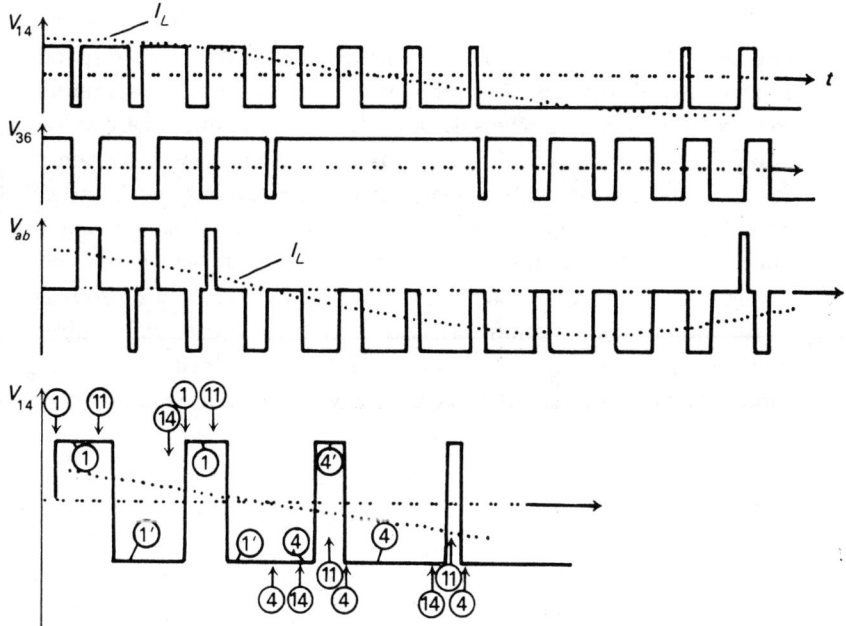

Fig. 179   Phase-potential waveforms V_{14} and V_{36}, and line–line voltage V_{ab} and line current I_L waveforms for the invertor of Fig. 178 with trapezoidal modulation

The change in firing sequence required by the reversal of line current is indicated at (d)

237

phase-potential waveform, it is apparent that, when 11 has turned off 1 in the modulated interval, the load current flows through 1' and the phase potential is at $-V_s$; 1 is then refired. While 1' is conducting, the commutating capacitor remains charged to $4V_s$ (with polarity $+$ $-$); and, if 1 were refired without first reversing this polarity, the subsequent turnoff process for 1 would not work. While 1' is conducting, the opportunity is therefore taken to reverse resonantly the polarity of C_{14} by firing 14.

When the phase current has reversed, conduction of 4 gives the phase potential $-V_s$; and 14, turning off 4, initiates the conduction of 4' giving a phase potential $+V_s$. Again the resonant reversal must be performed, now by firing 11 while 4' conducts. The reversal of phase current must therefore be monitored, to change over the firing times of 11 and 14.

From Fig. 178 and for the chosen value of $k_i = \frac{1}{2}$, the minimum effective time at the potential $+V_s$ for the phase voltage corresponds to 150° of resonant oscillation; the minimum time at $-V_s$ corresponds to 180°, the resonant reversal time for the capacitor. These values apply for load current in the direction of thyristor 1, and are reversed for the opposite direction of load current. As t_{RV} is typically 50 μs, eqn. 8.27 gives a half-resonant period of 75 μs, short enough to allow many phase-potential periods at $+V_s$ and $-V_s$ within one invertor period lasting 1667 μs for an output frequency of 100 Hz.

Individual commutation has also been used with variable direct voltage where the lower harmonic content (obtained by avoiding modulation) of the output voltage is of value (Espelage *et al.* 1969).

9 FIRING-CONTROL CIRCUITS

9.1 Introduction

The firing pulse, or the prolonged firing signal, has been taken for granted in the description of power convertors and controllers. The correct operation of any thyristor circuit is just as dependent on the absence of all unwanted, or spurious, gate signals as it is on the presence of correctly timed firing signals. The thyristor requires only about 0·25 V to fire the most sensitive devices of a batch, thus placing a considerable emphasis on noise free control circuits. The very speed and sensitivity of the thyristor makes the search for spurious signals difficult, as brief pulses of only a few microseconds can be sufficient to fire a sensitive device. It is appropriate, therefore, to make a few observations about the points at which power and control circuits meet, with the object of highlighting possible causes of spurious pulses and noise on control circuitry. This is followed by a discussion of the pulse amplifier which feeds the thyristor gate via an isolating pulse transformer. Two techniques of obtaining isolated and prolonged pulses with small transformers are described. Delay-angle control is required for rectifiers, a.c. regulators, synchronous invertors and cycloconvertors, and various techniques are reviewed. Choppers require frequency- and delay-angle control, and conveniently lead into the frequency- and sequence-control circuits required for capacitor-commutated invertors.

9.2 Power-control interface

Thyristors are now available to withstand a current rate at turnon of 300 A/μs, at which rate an inductance of 1 μH, involving only 1–2 m of cable, can develop 300 V between its ends. A mutual inductance of a similar value between a power cable and a control wire can clearly result in large voltage spikes in control wires, with a likelihood of generating a spurious firing pulse and radio interference (Von Zastrow). If high rates of change are used in power convertors, it is important that the cable layout of the power circuit should minimise the enclosed area of all high di/dt loops and that control wiring is well separated from these loops. Capacitive coupling can also be important since thyristors can now withstand safely voltage rates of 200 V/μs. However, if the power and control circuits are physically separated, the interwiring capacitance will give rise to only very small current transients which, in low impedance transistor circuitry, are of little trouble.

The most obvious and necessary interface is at the thyristor-gate pulse transformer. Electrical isolation between power and control circuits is universally applied to high-power convertors. It is usual to provide each thyristor with its own pulse transformer which, using ferrite core material, is typically of less volume than 1 in^3. The gate leads are screened, although, with careful routing, twisted pairs are acceptable. Transients of anode voltage in the off state can induce gate-voltage transients, so that it is clearly important that the transistor drive circuitry feeding the primary of the pulse transformer(s) should be insensitive to this noise. Problems have occurred with regenerative blocking-oscillator stages owing to the regenerative coupling between the pulse transformer and the transistor base.

The next obvious interface arises from the closed-loop control of the thyristor convertor, which requires the power current and sometimes the voltage to be measured for automatic limitation or control in the logic or analogue sense. Here 50 Hz transformers are likely to introduce more interwinding capacitance than the small gate pulse transformers, so that earth screening is often necessary between the power and control windings. High-frequency transistor choppers are useful when a d.c. power signal must be measured while retaining electrical isolation: the Hall plate also offers a convenient way of measuring current with electrical isolation and low capacitance.

A third interface arises since the control-circuit power supplies are energised from the same source which feeds the power convertor. It has already been shown how a power convertor generates harmonic currents which produce harmonic voltage drops across the supply reactance. If the control circuits are energised from the secondary of a rectifier transformer, the line–line voltage falls momentarily to zero at each commutation. Even the primary voltage has noticeable sudden dips of voltage at commutations. These commutation dips can occur at varying times in each halfcycle, including the beginning and the end, so that, where the supply waveform is used to synchronise the thyristor-firing pulses, and perhaps also to determine the delay angle, filtering must be inserted to smooth the distorted waveform (Stahl, 1969; Battersby, 1969).

The stray interface remains the most troublesome, largely because it is so ill defined. In circuit design it is necessary to be prepared for contact bounce on all forms of mechanical switch, and the possible consequence of high dV/dt on the second closure when snubbing capacitors have already been charged. Contactors may also form an interface, since auxiliary contacts for interlocking the control circuits are close to the power contacts. The swamping of stray capacitance with parallel capacitor(s) at the control-unit terminals generally removes noise from auxiliary contacts without introducing a significant time constant.

9.2 Gate-pulse stage

Thyristor manufacturers provide a spread of possible firing points in the V_g–I_g plane. Noise must be kept lower than that required to fire the most sensitive device, while the load line of the pulse circuit (from the open-circuit voltage to the short-circuit current) must pass above and to the right of the firing point of the least sensitive device at the lowest temperature (Fig. 180). For power applications where high thyristor di/dt occurs, an open-circuit voltage of 10–15 V and a short-circuit current of 2–4 A is required, with a risetime less than 1 μs. (Special thyristors have been developed which can safely be fired with gate currents an order of magnitude less, and with a slow risetime.) The large, fast gate pulses, if prolonged, are likely to overheat the gate, so that 2-level pulses are resorted to as shown in Fig. 181.

The self-commutated invertor poses the most difficult firing problem, as the commutation process requires a large, fast gate pulse to

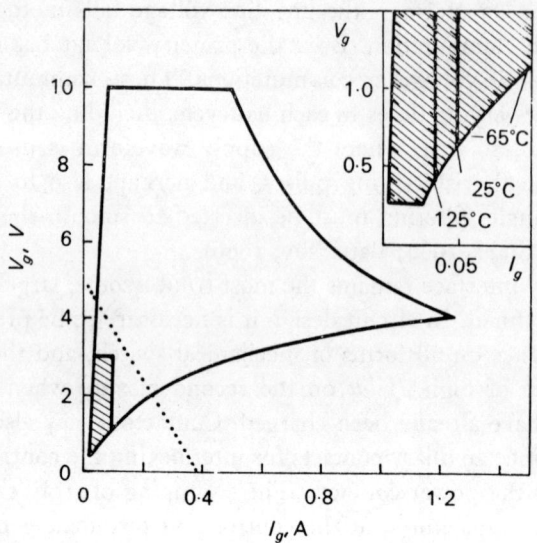

Fig. 180 Permitted area for instantaneous applied gate voltage and gate current for a typical spread of thyristor characteristics in a type batch

Shaded area represents the locus of all firing points, detailed in the inset to show the effect of junction temperature. A typical load line for a gate-pulse circuit is shown dotted

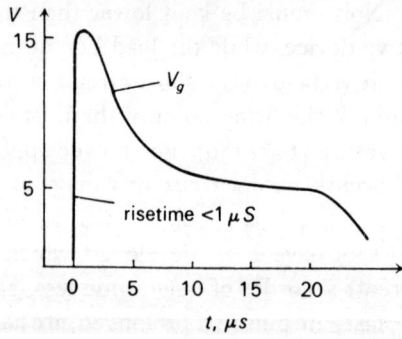

Fig. 181 2-level gate pulse which is desirable when the thyristor must accept a high *di/dt* at turnon

enhance the *di/dt* capability of the thyristors for fast commutation; yet the load current may not transfer to the thyristor for several milliseconds, thus demanding a very prolonged firing signal. One solution is to use a train of short pulses, each pulse having a duration of 10–20 μs and a repetition period of 40–100 μs. Commercial gate-drive units are available with pulse-train output. The other solution is to supply the pulse-transformer primary with a square wave which can be switched on and off rapidly for the required duration of gate signal, and to fullwave-rectify the secondary output voltage. On

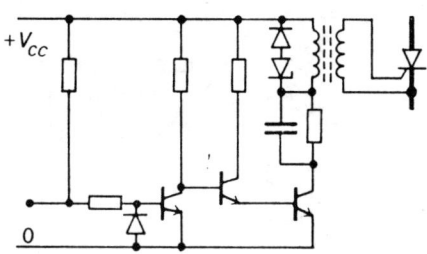

Fig. 182 Typical gate-pulse stage suitable for individual pulse
or pulse-train operation

Capacitor in parallel with the current-limiting resistor (which fixes the load line) provides the initially higher voltage for the 2-level pulse illustrated in Fig. 181

choosing a high square-wave frequency (say, 10 kHz), the 'pulse' transformer is no bigger than for single-pulse or pulse-train firing. The rectified square wave has the advantage of providing virtually continuous gate drive since the gaps between the rectified halfcycles are only 1–2 μs.

Discrete components are still used for the transistor output stage owing to the power rating involved, but integrated circuits are being adopted rapidly for the lower-power phase-control and logic-control functions described later.

A typical single-pulse or pulse-train output stage is shown in Fig. 182. Note the Zener diode in series with the freewheeling diode, to allow a substantial reverse voltage to appear across the primary to collapse rapidly the core flux between successive pulses of a pulse train. The series gate resistance, which fixes the load line, has been transferred to the primary, and, with the parallel capacitor, it provides the 2-level pulse. As the transformer has typically only 20 turns per winding, it is not difficult to achieve a high interwinding-voltage rating of about 2 kV, in spite of the small physical size.

For the square-wave output stage, a pair of transistors in push–pull drive the centre-tapped primary. On and off gating is achieved by a third transistor which biases the square-wave source feeding the output transistor bases (Fig. 183). [See also (Jarratt, 1963).]

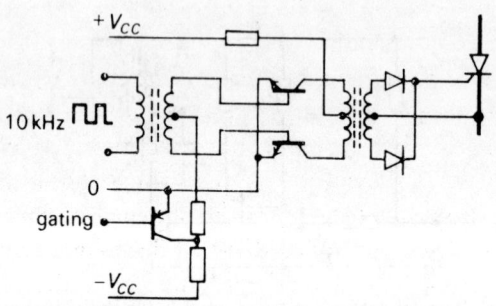

Fig. 183   Gate-pulse stage which produces virtually continuous
gate signal by fullwave rectification of square wave

The gating transistor, when on, allows the 10 kHz square wave to switch the output transistors off alternately. When the gating transistor is off, both output transistors conduct simultaneously, reducing the likelihood of spurious gate pulses by short circuiting the transformer primary

9.4 Delay-angle control

Delay-angle control is required for controlled rectifiers, a.c. regulators and cycloconvertors. 2-thyristor choppers also require delay-angle control to control the mark/space ratio, but fewer problems arise for this case.

The control of delay angle presupposes a supply from which the delay is measured, and, for nearly all applications, this is the main power supply to the conversion equipment itself. This supply voltage is always distorted to some extent by the commutations of the convertor, and it is essential that these distortions cause no maloperation of the delay control circuits. Several techniques have been developed, all using transistors or integrated circuits, but each solving the interference problem in a slightly different way.

The first uses principles established for the firing-angle control of mercury-arc rectifiers. The crossover point between a sine wave derived from the supply and a variable d.c. level whose value determines the delay is detected and used to initiate the gate firing pulse. The sine wave is obtained by filtering the a.c. supply with high-stability RC components. The phase shift of the several filters for a polyphase rectifier must be kept within a close tolerance, less than $1°$, to avoid components of direct current in the rectifier transformer or supply. It is imperative that the firing pulses do not disappear should there be a transient drop in the amplitude of the sine wave or a rise in the d.c. level. To ensure that the gate pulses are always present, the positive and negative crests of the sine wave have pulses added to them, as shown in Fig. 184, which shows that a range of pulse movement close to $180°$ can be obtained by this method. The same direct-voltage level is used for all delay-angle control circuits, but each thyristor (or string or group acting together) has an appropriately phased sine wave to allow delay-angle control from o to $180°$ for the particular thyristor(s). The level detection can be performed using a Schmitt trigger, or even a transistor with its base–emitter voltage temperature-compensated with a diode according to Fig. 185, which also shows the typical arrangement for the remainder of the circuitry (Battersby, 1969).

The use of a sine wave for delay-angle control gives a conveniently linear relationship between the d.c. control signal and the output of a controlled rectifier.

The second technique does not use the filtered supply, but generates another waveform, usually a ramp, in synchronism with the supply. The unijunction transistor (u.j.t.) is a particularly useful device for this purpose (Sylvan; Gutzwiller, 1967), as it applies negligible loading to the RC circuit which generates the ramp, yet can pass a discharge current of typically 1 A when the capacitor voltage exceeds the triggering level of the u.j.t. Fig. 186 shows such a circuit, in which the u.j.t. fires typically when its emitter reaches $+6$ V, producing a 6 V pulse as C discharges into R_2 and the follow-

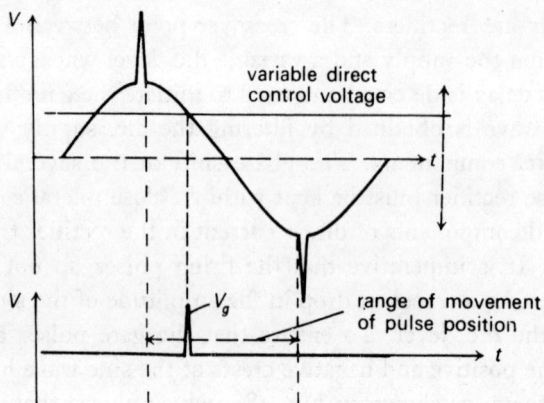

Fig. 184 Delay-angle control using the crossover between a sine-wave
and a variable direct control voltage

Pulses at the sine-wave crest are added to ensure that the crossover is never lost
should the sine-wave amplitude fall when the control voltage is at its maximum
values, calling for zero or 180° delay

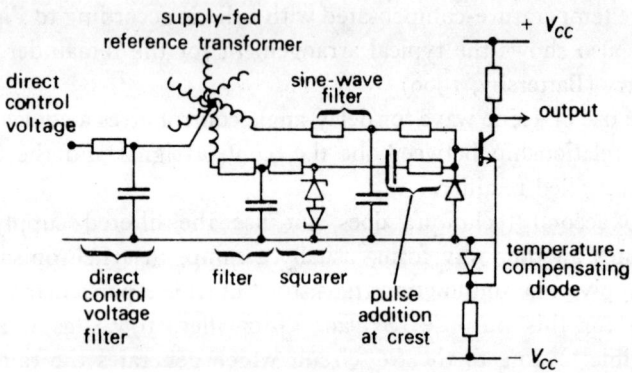

Fig. 185 Typical delay-angle circuit operating on the
principle of Fig. 184

The sine-wave filter must produce a 30° phase shift, and the other filter a 60°
phase shift so that the pulse addition occurs at the crest

246

ing circuitry. The rate of charging of C is varied with Tr_1 from a minimum value set by R_3 to give a delay angle of $175°$, say, to a maximum value which gives a delay angle approaching zero. The circuit only works during negative halfcycles (thus providing the synchronisation with the supply) since, for positive halfcycles, Tr_2 short-circuits C.

The circuit is less prone to commutation-dip interference except where this affects the zero crossover point for the supply waveform. Even this influence can be removed by replacing the base resistance

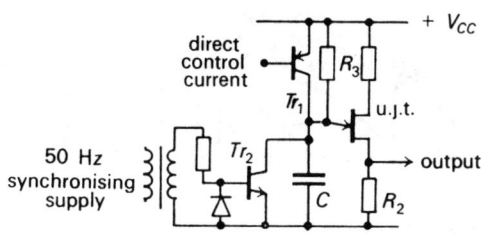

Fig. 186 Delay-angle-control circuit which generates a sawtooth waveform during alternate halfcycles of the supply voltage, from which variable-delay output pulses are generated

for Tr_2 with an inductor, which acts as a lowpass filter and provides a phase shift of approaching 90° for the base current of Tr_2; the phase shift is then much less affected by frequency than for the previous circuit.

Fixed-frequency chopper circuits also require delay-angle control, since the load thyristor is fired at a variable delay after firing the turnoff thyristor. The chopper frequency is determined by an oscillator. A relaxation-type oscillator, generating a sawtooth waveform, is most convenient for delay-angle control, as the sawtooth is readily compared with a variable d.c. level to obtain a variable delay pulse, as shown in Fig. 187 (Rayworth, 1965).

9.5 Frequency generation for invertors and choppers

Capacitor-commutated invertors operate at a frequency generated by the control circuits. Only for parallel operation or for synchronisation to an existing supply is the firing frequency governed by conditions outside the control circuits. The heart of the frequency-generation circuits is thus an oscillator, ranging from a u.j.t. relaxation oscillator (Fig. 186, with 50 Hz synchronisation omitted) having a frequency stability within 1 % typically, to a crystal-controlled system capable of a stability within 1 part in 10^8.

Fixed-frequency invertors pose few frequency-generator problems, but a variable-frequency output requires an oscillator which can sweep the entire frequency range in response to an analogue control

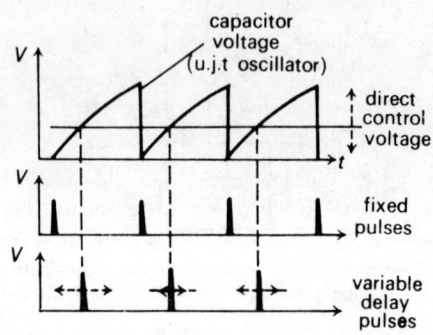

Fig. 187   Principle of delay-angle control for a chopper

The top waveform shows the comparison of a sawtooth oscillator output and a variable direct control voltage. The lower two waveforms show the nondelayed and delayed pulses

signal. A typical range of output frequency is 1–100 Hz, which requires an oscillator adjustable in the range 6–600 Hz for a 3-phase bridge invertor. A frequency stability within 1 % of the output frequency is difficult to attain using the simple u.j.t. oscillator when the output frequency is at its lower values, but is readily obtained if the frequency stability is specified as a percentage of the maximum frequency. (The latter is common practice when expressing speed stability of a variable-speed drive.) Coverage by means of switched ranges is not acceptable for motor drives, as the motor draws excessive currents at higher than the rated slip frequency, and it is not economical to design the invertor for these high currents. Analogue oscillators can obtain a frequency stability of within 0·05 % over a range 60–600 Hz using temperature control and precision components, but this represents the limit of analogue techniques. For greater stability, a crystal standard must be used, coupled to a variable-ratio divider to obtain the variable output frequency; an infinitely variable output frequency cannot be so achieved. With a 1 MHz oscillator and a maximum output frequency of $10^6/1024$ Hz, say, the largest step change in frequency occurs when the 15-bit counter fed from the oscillator is changed from a 1023 to 1024 count, a change of less than 0·1%. The lowest output frequency occurs when the whole of the 15-bit counter is used, giving an output frequency of 30·5 Hz, a frequency range of 30:1, and progressively closer resolution between steps as the output frequency is reduced.

It has been suggested that the simpler single-phase invertors can be fed directly from an astable multivibrator: this is not considered good practice since there are different timing components for each halfcycle. Unequal halfcycles produce core saturation when the invertor load includes transformers or where the invertor has its own output transformer.

9.6 Sequence-control circuits

Sequence-control circuits are required to produce a sequence of firing signals of the correct timing and duration to initiate the desired conduction pattern for the thyristors (Bradley *et al.* 1964). The simplest sequence-control circuit is bistable, into which frequency-generator pulses are fed, and from which each of two outputs produces interlaced signals in synchronism with every alternate frequency-generator pulse. When used to fire the thyristors of a single-phase invertor, the bistable (divide-by-two) ensures exact equality of length for the alternate halfcycles, avoiding transformer saturation.

For a 3-phase bridge invertor, a ring-of-six counter (acting like a car distributor) is required to provide correctly timed firing signals for the invertor thyristors, and, where auxiliary commutation is used, e.g. half commutation, a bistable is also required to distribute pulses alternately to the commutation thyristors. Where circulating currents must be allowed to decay before firing the thyristors, appropriate delays must be included in the control circuits.

A modulated-wave invertor requires the combination of sequence control and mark/space-ratio control. The latter is generally obtained by detecting the crossovers between a high-frequency triangular or sawtooth wave at the chopping frequency, and a lower-frequency modulating signal, often of trapezoidal waveform. Fig. 188 illustrates how these waveforms can generate the appropriate firing signals for modulation purposes.

When the invertor must operate at a variable frequency, some signal durations must vary inversely with the frequency, and their timing must be determined solely from the ring counter. When a signal between two frequency-generator pulses is required to occur after

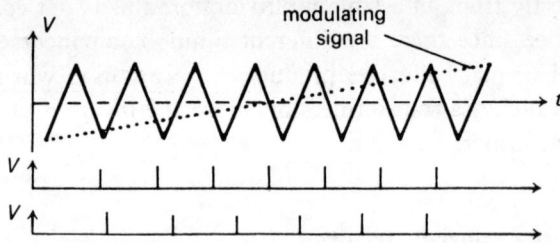

Fig. 188 Control pulses for waveform modulation produced by detecting the crossovers between a triangular (or sawtooth) waveform at the desired frequency of mark/space-ratio control, and a lower frequency modulating signal

a fixed proportion of the (variable) invertor period, an intervening ramp of fixed amplitude, synchronised to the ring counter, can be used. A suitable ramp signal already exists within a sawtooth frequency generator (e.g. u.j.t. oscillator) which, with appropriate gating logic, yields a signal which can be positioned anywhere within a particular invertor period regardless of the frequency. This is shown in Fig. 189.

The above comments by no means describe all the techniques available or useful for sequence control, but are intended simply to show how appropriately timed signals may be obtained.

9.7 Closed-loop controls

Variable-speed drives, in particular, require closed-loop controls or closed-loop limit circuits to protect the thyristor convertor against excessive currents, and to provide the necessary accuracy of control. The general principles of closed-loop control, including stability analysis and compensation techniques, are well established (Fallside and Wortley, 1969; Jackson, 1969; Lipo and Krause, 1969).

One aspect of closed-loop control requires a more detailed treatment here because it is likely to apply to most convertors in control loops.

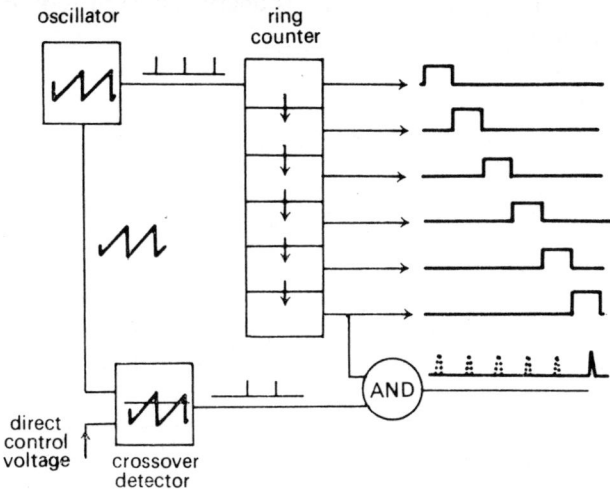

Fig. 189 Typical sequence-control techniques for a 3-phase bridge invertor, in which a pulse can be positioned anywhere within a particular invertor period via the crossover detector and AND gate

9.7.1 Ripple instability

The thyristor convertor is capable of very rapid adjustment of its output as there is virtually no 'inertia' between the analogue control signal representing the desired output and the delay angle which fixes the actual output. If the remainder of the system comprising the closed loop also has a fast response, the possibility exists for an

251

instability not predicted by conventional control theory (Fallside and Farmer, 1967; Fallside, 1968). The instability can arise because of the 'sample-hold' characteristic of the thyristor convertor. Once a firing signal has been delivered, the output voltage for the next repetition period is fixed and is not controllable. If one imagines one repetition period having too great an output voltage, the closed-loop controls will ensure that the next period has too small an output voltage, to obtain the correct mean value. A stable limit cycle can exist at a subharmonic of the ripple frequency, so that the output-voltage waveform and firing-angle delay are modulated at this subharmonic frequency. This phenomenon has been investigated and analysed, and the occurrence of a subharmonic oscillation can now be predicted.

BIBLIOGRAPHY

ADAMSON, C., and HINGORANI, N. G. (1960): 'High voltage direct current power transmissions' (Garraway)

AINSWORTH, J. D. (1967): 'Harmonic instability between controlled static convertors and a.c. networks', *Proc. IEE*, **114**, (7), pp. 949–957

AMATO, C. J. (1966): 'An a.c. equivalent circuit for a cycloconvertor', *IEEE Trans.*, **IGA-2**, pp. 358–362

ARRILLAGA, J., and GALANOS, G. (1969): 'Fault-development control in a.c.– d.c. convertors', *Proc. IEE*, **116**, (7), pp. 1201–1208

BATTERSBY, C. F. (1969): 'Present techniques in gate firing', *IEE Conf. Publ.* 53, Pt. 1, pp. 146–153

BEASLEY, J., and WHITE, G. (1965): 'The use of thyristors for the control of a d.c. traction motor operating from a 600 V line supply', *ibid.*, Pt. 1, pp. 187–195

BEDFORD, B. D., and HOFT, R. G. (1964): 'Principles of inverter circuits' (Wiley)

BERGMAN, G. D., and KNOTT, R. D. (1965): 'Some design aspects of silicon 5-layer symmetrical switches', *IEE Conf. Publ.* 17, Pt. 1, pp. 26–36

BIRD, B. M., MARSH, J. F., and MCLELLAN, P. R. (1969): 'Harmonic reduction in multiplex convertors by triple-frequency current injection', *Proc. IEE*, **116**, (10), pp. 1730–1734

BLAND, R. J. (1967): 'Factors affecting the operation of a phase-controlled cycloconvertor', *ibid.*, **114**, (12), pp. 1908–1916

BLUNDELL, A. J., GARSIDE, A. E., HIBBERD, R. G., and WILLIAMS, I. (1961): 'Silicon power rectifiers', *ibid.*, **108**A, pp. 273–288

BORST, D. W., DIEBOLD, E. J., and PARRISH, F. W. (1966): 'Voltage control by means of power thyristors', *IEEE Trans.*, **IGA-2**, pp. 102–124

BOWLER, P. (1965): 'The application of a cycloconvertor to the control of induction motors', *IEE Conf. Publ.* 17, Pt. 1, pp. 137–145; Pt. 2, pp. 42–43

BRADLEY, D. A., CLARKE, C. D., DAVIS, R. M., and JONES, D. A. (1964): 'Adjustable-frequency invertors and their application to variable-speed drives', *Proc. IEE*, **111**, (11), pp. 1833–1846

BRISBY, K. (1965): 'High-power thyristor invertors for standby a.c. power supplies', *IEE Conf. Publ.* 17, Pt. 1, pp. 168–177

BROWNSEY, C. M., and CSUROS, L. (1963): 'Harmonic distortion due to rectifier loads on a.c. supply systems', *IEE Conf. Publ.* 8, pp. 141–147

BS 4417 (1969): Semiconductor rectifier equipments, pp. 38–39

BULL, J. H. (1968): 'Voltage spikes in l.v. distribution systems and their effects on the use of electronic control equipment', ERA report 5254

CHIRGWIN, K. M. (1965): 'A variable-speed constant-frequency generating system for a supersonic transport', *IEEE Trans.*, **AES-3**, pp. 387–393

CORBYN, D. B. (1965): 'Voltage surge control in thyristor equipment', *IEE Conf. Publ.* 17, Pt. 1, pp. 89–101

CORBYN, D. B., and POTTER, N. L. (1960): 'The characteristics and protection of semiconductor rectifiers', *Proc. IEE*, **107**A, pp. 255–269

CORDINGLEY, B. V. (1969): 'A method of studying thyristor operating phenomena using recombination radiation', *IEE Conf. Publ.* 53, Pt. 1, pp. 8–13

CORY, B. J. (Ed.) (1955): 'High voltage direct current convertors and systems' (Macdonald)

DAVIS, R. M. (1969): 'Thyristor control of a multiload system with d.c. supply', *Proc. IEE*, **116**, (5), pp. 801–810

DEWAN, S. B., and HAVAS, G. (1969): 'A.C.–a.c. frequency convertors for induction heating and melting', *IEE Conf. Publ.* 53, Pt. 1, pp. 440–447

DIEBOLD, E. J. (Ed.) (1962): 'The controlled rectifier' (International Rectifier Corporation)

DORTORT, I. K. (1953): 'Extended regulation curves for 6-phase double way and double wye rectifiers', *Trans. Amer. Inst. Elect. Engrs.*, **72**, Pt. I, pp. 192–202

DORTORT, I. K. (1958): 'Current balancing reactors for semiconductor rectifiers', *ibid.*, **77**, Pt. I, pp. 452–456

DUFF, D. L., and LUDBROOK, A. (1968): 'Semiconvertor (half-controlled) rectifiers go high power', *IEEE Trans.*, **IGA-4**, pp. 185–192

DURNYA, F. (1968): 'Selection of s.c.r.s for single-phase d.c. motor drives', International Rectifier Corporation *Rectifier News*, summer

ESPELAGE, P. M., CHIERA, J. A., and TURNBULL, F. G. (1969): 'A wide range static invertor suitable for a.c. induction motor drives', *IEEE Trans.*, **IGA-4**, pp. 438–445

EVANS, R. D., and MULLER, H. N. (1939): 'Harmonics in the a.c. circuits of grid-controlled rectifiers and invertors', *Trans. Amer. Inst. Elect. Engrs.*, **58**, pp. 861–870

FALLSIDE, F. (1968): 'Ripple instability in closed-loop pulse-modulation systems including invertor drives', *Proc. IEE*, **115**, (1), pp. 218–228

FALLSIDE, F., and FARMER, A. R. (1967): 'Ripple instability in closed-loop control systems with thyristor amplifiers', *ibid.*, **114**, (1), pp. 139–152

FALLSIDE, F., and WORTLEY, A. T. (1969): 'Steady-state oscillation and stabilisation of variable-frequency invertor-fed induction-motor drives', *ibid.*, **116**, (6), pp. 991–999

FEINBERG, R., and CHEN, W. Y. (1964a): 'Commutation phenomena in a static power convertor', *ibid.*, **III**, (1), pp. 125–134

FEINBERG, R., and CHEN, W. Y. (1964b): 'Commutation reactance of the transformer in a static power convertor', *ibid.*, **III**, (1), pp. 135–142

FELTBOWER, B. (1969): 'Thyristor contactors', *IEE Conf. Publ.* 53, Pt. 1, pp. 448–464

FRERIS, L. L. (1961): 'The universal characteristic of the 3-phase bridge converter', *Direct Current*, **6**, pp. 198–201

FRERIS, L. L. (1966): 'Analysis of a hybrid bridge rectifier', *ibid.*, **11**, pp. 22–33

FRERIS, L. L. (1967): 'Effects of interaction among groups in a multigroup a.c.–d.c. convertor', *Proc. IEE*, **114**, (7), pp. 965–973

G 5/2 (1967): 'Supplies to convertor equipments: harmonic distortion and permissible pulse number of consumers' rectifiers and invertors', Electricity Council Engineering Recommendation

GALLOWAY, J. H.: 'Using the triac for control of a.c. power', General Electric Co. application note 200.35

GARDNER, G. E., and FAIRMANER, D. (1968): 'Alternative convertor for h.v. d.c. transmission', *Proc. IEE*, **115**, (9), pp. 1289–1296

GENTRY, F. E. (1958): 'Forward current surge failure in semiconductor rectifiers', *Trans. Amer. Inst. Elect. Engrs.*, **77**, Pt. 1, pp. 746–750

GENTRY, F. E., GUTZWILLER, F. W., HOLONYAK, N., JUN., and VON ZASTROW, E. E. (1964): 'Semiconductor controlled rectifiers' (Prentice Hall)

GERECKE, E. (1950): 'Some considerations on the voltage distortions caused in 3-phase networks by higher harmonics', CIGRÉ, Paris, Paper 320

GOLDEN, F. B. (1968): 'Power semiconductor I^2t ratings', Proceedings of the IEEE international convention, New York, p. 147

GUTZWILLER, F. W.: 'An introduction to the controlled avalanche silicon rectifier', General Electric Co. application note 200.27

GUTZWILLER, F. W. (Ed.) (1967): 'S.C.R. manual' (General Electric Co., 4th edn.)

HALL, J. K. (1969): 'Forced commutation of thyristors connected in series strings', *IEE Conf. Publ.* 53, Pt. 1, pp. 365–371

HAMILTON, R. A., and LEZAN, G. R. (1967): 'Thyristor adjustable frequency power supplies for hot strip mill run-out tables', *IEEE Trans.*, **IGA-3**, pp. 168–175

HECK, R., and MEYER, M. (1963): 'A static-frequency-changer-fed squirrel-cage motor drive for variable speed and reversing', *Siemens Rev.*, **11**, pp. 401–409

HEUMANN, K. (1969): 'Development of inverters with forced commutation for a.c. motor speed control up to the megawatt range', *IEEE Trans.*, **IGA-5**, pp. 61–67

HEY, J. C.: 'Series operation of s.c.r.s', General Electric Co. application note 200.40

HINGORANI, N. G., and HALL, J. K. (1965): 'Use of capacitors for reduction of commutation angle in static high-power convertors', *Proc. IEE*, **112**, (12), pp. 2333–2341

HÖLTERS, F. (1961): 'Current and voltage conditions from no-load to short circuit in 3-phase bridge circuits', *Direct Current*, **5**, pp. 112–122

HUMPHREY, A. J. (1968): 'Inverter commutation circuits', *IEEE Trans.*, **IGA-4**, pp. 104–110

IEC 146 (1963): Monocrystalline rectifier cells, stacks, assemblies and equipments, p. 87

IEE Conf. Publ. 17 (1965): 'Power applications of controllable semiconductor devices'

IEE Conf. Publ. 53 (1969): 'Power thyristors and their applications'

IEEE Conf. Rec. (1965): 'Industrial static power conversion conference record', Philadelphia

ITO, N., and SEKINE, S. (1968): 'Computer aided control of h.v. d.c. power systems', *Elect. Engng. Japan*, **88**, pp. 50–59

JACKS, E. (1969): 'High-speed fuse protection for silicon diodes and thyristors', *IEE Conf. Publ.* 53, Pt. 1, pp. 116–124

JACKSON, R. D. (1969): 'Oscillations in single convertor drives', *Proc. IEE*, **116**, (4), pp. 633–638

JARRATT, T. J. (1963): 'Transistorised s.c.r. firing circuits', *Mullard Tech. Commun.*, **65**, pp. 141–157

JONES, D. (1961): 'Variable pulse width invertor', *Electron. Equipment Engng.*, Nov. 1961, pp. 29–30

JURI, I., and YOSHIDA, T. (1969): 'Thyristoformer for 100 kA', *IEE Conf. Publ.* 53, Pt. 1, pp. 234–240

KALIS, H., and LEMMRICH, J. (1969): 'D.C. chopper with high switching reliability and without limitation of the adjustable mark/space ratio', *ibid.*, Pt. 1, pp. 208–215

KING, K. G. (1965): 'Thyristor a.c. power regulators', *IEE Conf. Publ.* 17, Pt. 1, pp. 206–210

KNIGHT, H. DE B. (1960): 'The arc discharge', pp. 283–294 (Chapman & Hall)

KNOTT, R. D., WADHAM, E., and FOSTER, A. (1965): 'The influence of the surface on the high-voltage performance of thyristors, and techniques to eliminate surface effects', *IEE Conf. Publ.* 17, Pt. 1, pp. 56–65

KUSKO, A., and SZPAKOWSKI, B. (1965): 'Load ranges of series s.c.r. invertors', *Electro-Technology*, April, pp. 76–80

LANDAU, I. D. (1969): 'Wide range speed control of 3-phase squirrel-cage induction motors using static frequency convertors', *IEEE Trans.*, **IGA-5**, pp. 53–60

LAWN, F. (1962): 'Static inverters using s.c.r. with pulse-width control', Proceedings of the 16th annual power sources conference, Fort Monmouth, NJ, pp. 154–156

LAWSON, L. J. (1968): 'The practical cycloconvertor', *IEEE Trans.*, **IGA-4**, pp. 141–144

LERSTRUP, K. (1965): 'High-speed fuses for the protection of diodes and thyristors', *IEE Conf. Publ.* 17, Pt. 1, pp. 111–120

LI, K.Y.G. (1968): 'New 3-phase invertor circuit', *Proc. IEE*, **115**, (11), pp. 1677–1683

LIGHT, W. R., JUN., and MCVEY, E. S. (1967): 'A synchronous tap-charger applied to step-up cycloconvertors', *IEEE Trans.*, **IGA-3**, pp. 244–249

LIPO, T. A., and KRAUSE, P. C. (1969): 'Stability analysis of a rectifier-invertor-induction motor drive', *ibid.*, **PAS-88**, pp. 55–66

LLOYD, S. (1969): 'A thyristor a.c. regulator with sinusoidal output', *IEE Conf. Publ.* 53, Pt. 1, pp. 168–176

LUDBROOK, A., and MURRAY, R. M. (1965): 'A simplified technique for analysing the 3-phase bridge rectifier circuit', *IEEE Trans.*, **IGA-1**, pp. 182–186

LYALL, A. M. (1969): 'Thyristor control by cycle selection or phase angle firing on a 25 kV 50 Hz traction unit', *IEE Conf. Publ.* 53, Pt. 1, pp. 531–537

MAPHAM, N. (1963): 'The rating of s.c.r.s when switching into high currents', *IEEE Trans.*, **83**, Commun. & Electron., pp. 515–519

MAPHAM, N. (1967a): 'An s.c.r. inverter with good regulation and sine-wave output', *ibid.*, **IGA-3**, pp. 176–187

MAPHAM, N. (1967b): 'Low cost ultrasonic frequency inverter using single s.c.r.', *ibid.*, **IGA-3**, pp. 378–388

MARTI, O. K., and WINOGRAD, H. (1930): 'Mercury arc power rectifiers' (McGraw-Hill)

MARTIN, A. M.: 'Development in selenium rectifiers', International Rectifier Co. (GB) application note GBAN-S-1

MCBREEN, J. P. (1960): 'Some considerations in the application of power rectifiers and convertors', *Proc. IEE*, **107**A, pp. 445–460

MCCOLL, J. D., and WHITAKER, P. (1965): 'Service experience with thyristors', *IEE Conf. Publ.* 17, Pt. 1, pp. 102–110

MCMURRAY, W. (1963): 'S.C.R. d.c.–d.c. power convertors', *IEEE Trans.*, **83**, Commun. & Electron., pp. 198–203

MCMURRAY, W. (1964): 'S.C.R. inverter commutated by an auxiliary impulse', *IEEE Trans.*, **83**, Commun. & Electron., pp. 824–829

MCMURRAY, W., and SHATTUCK, D. P. (1961): 'A silicon controlled rectifier invertor with improved commutation', *Trans. Amer. Inst. Elect. Engrs.*, **80**, Pt. 1, pp. 531–542

257

MCTAGGART, J. (1968): 'Analogue-computer study of a convertor jump phenomenon', *Proc. IEE*, **115**, (8), pp. 1173–1177

MELLGREN, G. (1965): 'Thyristor convertors for motor drives: some experience in design and operation', *IEE Conf. Publ.* 17, Pt.1, pp. 230–249; especially Fig. 6, p. 241

MOKRYTZKI, B. (1967): 'Pulse-width modulated inverters for a.c. motor drives', *IEEE Trans.*, **IGA-3**, pp. 493–503

MORGAN, R. E. (1963): 'Time ratio control with combined s.c.r. and saturable reactor commutation', *ibid.*, **83**, Commun. & Electron., pp. 366–373

MORI, H. (1969): 'D.C. 200 V, 330 kA thyristor rectifier for electrolysis', *IEE Conf. Publ.* 53, Pt. 1, pp. 225–233

MURPHY, R. H., and NAMBIAR, K. P. P. (1961): 'A design basis for silicon controlled rectifier parallel invertors', *Proc. IEE*, **108** B, pp. 556–562

MURRAY, R., JUN. (Ed.) (1963): 'Thyristor manual' (Westinghouse Electric Corporation, 1st edn.)

NEWBERY, P. G. (1969): 'The correct protection of power thyristors by high-speed h.r.c. fuselinks', *IEE Conf. Publ.* 53, Pt. 1, pp. 125–131

NEWSAM, B. U. (1969): 'Transient thermal resistance: its measurement and use in the rating of thyristors', *ibid.*, Pt. 1, pp. 85–93

OHNO, E., and AKAMATSU, M. (1966): 'Variable frequency s.c.r. inverter with an auxiliary commutating circuit', *IEEE Trans.*, **MAG-2**, pp. 25–30

ONODA, Y., IZAWA, S., and KAWAKAMI, N. (1969): 'Thyristor applications to electric rolling stock', *ibid.*, **IGA-5**, pp. 141–148

OTSUKA, M. (1969): 'A new edge contour for Si high-voltage thyristors', *IEE Conf. Publ.* 53, Pt. 1, pp. 32–38

OTT, R. R. (1963): 'A filter for s.c.r. commutation and harmonic attenuation in high power inverters', *IEEE Trans.*, **82**, Commun. & Electron., pp. 259–262

PAYNE, R. A., and REEVES, E. S. (1963): 'Switch-off circuits for s.c.r.s operating on d.c.', *Mullard Tech. Commun.*, **65**, pp. 158–161

POLLARD, E. M., FLAIRTY, C. W., HODGES, M. E., and LAUKAITIS, J. A. (1969): 'A 20 MW thyristor a.c. switch for induction heating power control and protection', *IEE Conf. Publ.* 53, Pt. 1, pp. 177–184

PUCHLOWSKI, K. P. (1945): 'Voltage and current relations for controlled rectification with inductive and generative loads', *Trans. Amer. Inst. Elect. Engrs.*, **64**, pp. 255–260

RAYWORTH, G. (1965): 'Variable phase s.c.r. trigger circuit', *IEE Conf. Publ.* 17, Pt. 1, pp. 121–123

READ, J. C. (1945): 'The calculation of rectifier and invertor performance characteristics', *Proc. IEE*, **92**, Pt. II, pp. 495–509

REEVE, J. (1967): 'Logic behaviour of h.v. d.c. convertors during normal and abnormal conditions', *ibid.*, **114**, (12), pp. 1937–1946

REEVE, J., and BURDETT, G. E. (1969): 'Accelerated recovery from commutation faults in bridge connected a.c.–d.c. convertors', *IEE Conf. Publ.* 53, Pt. I, pp. 518–522

RICE, J. B., and NICKELS, L. E. (1968): 'Commutation dV/dt effects in thyristor 3-phase bridge convertors', *IEEE Trans.*, **IGA-4**, pp. 665–672

RISSIK, H. (1935): 'Mercury arc current convertors' (Pitman)

ROBERTS, M. E., and ASHMAN, W. G. (1969): 'A thyristor-assisted mechanical onload tap changer', *IEE Conf. Publ.* 53, Pt. I, pp. 185–192

SCHAEFFER, J. (1965): 'Rectifier circuits theory and design' (Wiley)

SCHMIDT, A., JUN. (1958): 'Power factor of rectifiers', *Trans. Amer. Inst. Elect. Engrs.*, **77**, Pt. II, pp. 53–58

SHEPHERD, W. (1965): 'Steady state analysis of the series R–L circuit controlled by silicon controlled rectifiers', *IEEE Trans.*, **IGA-1**, pp. 259–265

SHEPHERD, W. (1966): 'Steady state analysis of single-phase, parallel R–L circuits controlled by s.c.r. pairs', *ibid.*, **IGA-2**, pp. 469–473

SHEPHERD, W. (1968): 'On the analysis of the 3-phase induction motor with voltage control by thyristor switching', *ibid.*, **IGA-4**, pp. 304–311

SOMOS, I., and PICCONE, D. E. (1969): 'Some observations of static and dynamic plasma spread in conventional and new power thyristors', *IEE Conf. Publ.* 53, Pt. I, pp. 1–7

SPREADBURY, F. G. (1962): 'Electronic rectification' (Constable)

STAHL, B. P. (1969): 'Interaction between s.c.r. drives', *IEEE Trans.*, **IGA-4**, pp. 596–599

STORR-BEST, J. L. (1965): 'Thyristor control of fluorescent lighting banks', *IEE Conf. Publ.* 17, Pt. I, pp. 178–185

SYLVAN, T. P.: 'The unijunction transistor: characteristics and applications', General Electric Co. application note 90.10

THOMPSON, R. (1963): 'Designing series s.c.r. inverters', *Electron. Des.*, 7 June, pp. 52–58; 21 June, pp. 62–65; 5 July, pp. 48–53

THOMPSON, R. (1968): 'A thyristor alternating voltage regulator', *IEEE Trans.*, **IGA-4**, pp. 162–170

TOTH, J. R., SCHOEFFLER, J. D., CHIRGWIN, K. M. (1963): 'Artificial commutation of static convertors', *ibid.*, **82**, Commun. & Electron., pp. 83–94

TURNBULL, F. G. (1963): 'Selected harmonic reduction in static d.c.–a.c. inverters', *ibid.*, **83**, Commun. & Electron., pp. 374–378

UHLMANN, E. (1955): 'Alternating voltage, direct-voltage regulation and power factor of convertor stations operating on a.c. systems of finite short-circuit capacity', *Proc. IEE*, **102**C, pp. 284–289

VEDDER, E. H., and PUCHLOWSKI, K. P. (1943): 'Theory of rectifier d.c. motor drive', *Trans. Amer. Inst. Elect. Engrs.*, **62**, pp. 863–869

VON ZASTROW, E. E.: 'Controlled rectifiers and radio interference', General Electric Co. application note 200.3

WALLACH, Y., ERLICKI, M. S., and BEN URI, J. (1963): 'Multiphase rectifier currents', *Proc. IEE*, **110**, (8), pp. 1434–1440

WARBURTON, W., LOOTENS, W. F., and STAVISKI, J. (1966): 'Pressure contact semiconductor devices', *IEEE Trans.*, **IGA-2**, pp. 474–479

WILLIAMSON, K. H. (1969): 'Simple 3-phase a.c. motor control system for motors below 5 h.p.', *IEE Conf. Publ.* **53**, Pt. 1, pp. 320–327

YAIR, A., ALPERT, W., and BEN URI, J. (1969): 'Bridge rectifiers with double and multiple supply', *Proc. IEE*, **116**, (5), pp. 811–821

YANAI, Y. (1965): 'The epitaxial controlled rectifier', International Rectifier Corporation *Rectifier News*, spring

Supplementary bibliography for second reprinting

This list, for the period 1970–1978, is limited to References in English only. It is not exhaustive, but gives an adequate coverage of recent work in the UK and USA. The References are grouped under the appropriate section of the main text.

1.2
GHANDI, S.K.: 'Semiconductor power devices' (Wiley Interscience, 1977)

1.3
CORMICK, J.A.F., and RAMSBOTTOM, M.J.: 'Behaviour of thyristors when turned on by gate current', *Proc. IEE*, 1976, 123, (12), pp. 1365–7
MONTEITH, W., and BEATTIE, W.C.: 'Modelling approach to the design of thyristor systems', *Proc. IEE*, 1975, 122, (6), pp. 625-9
HO, H.H.: 'Improved logic model for thyristors', *Proc. IEE*, 1974, 121, (5), pp. 345–7
BEATTIE, W.C., MONTEITH, W., and PANKER, J.H.: 'Analogue modelling of a thyristor', *Proc. IEE*, 1973, 120, (7), pp. 786–8
BEATTIE, W.C., and MONTEITH, W.: 'Digital modelling of a thyristor', *Proc. IEE*, 1973, 120, (7), p. 789
HTSUI, J.S.C., and SHEPHERD, W.: 'Method of digital computation of thyristor switching circuits', *Proc. IEE*, 1971, 118, (8), pp. 993–8
RICE, J.B.: 'Design of snubber circuits for thyristor converters', *IEEE Conf. Proc.*, 1969, pp. 485–489
MCMURRAY, W.: 'Optimum snubbers for power semiconductors'. IEEE Conf. 6th Annual Meeting, IA, 1971, pp. 885–93
GHANDI, S.K.: 'Semiconductor power devices' (Wiley Interscience.)
OSBORN, E.A.: 'Thyristors 1972 – state of the art review', *Electronics & Power*, 1972, (4), pp. 125–7
HARNDEN, J.D.: 'Power semiconductors: looking ahead', *IEEE Spectrum*, 1977, (8), pp. 40–45
BALENOVICH, J.D., GILLOTT, D.M., and MOTTO, J.W.: 'Thyristor high frequency ratings by concurrent testing and computer simulation', *IEEE Trans.*, 1973, IA, (3/4), pp. 227–35
OKA, H., and GAMO, H.: 'Electrical characteristics of high voltage high power fast switching reverse-conducting thyristor and its application for chopper use', *IEEE Trans.*, 1973, IA, (3/4), pp. 236–47

OGAWA, T., KAMEI, T., and MORITA, K.: 'Electrical characteristics of ultra high voltage thyristors and related problems', *ibid.*, (1/2), pp. 112−5

1.4
WILLIAMS, B.W.: 'State space computer triac model', *Proc. IEE,* 1978, 125, (5), pp. 413−5
SAMPLE, S.B., SCHEUER, P.R., and SILVA, L.F.: 'Reliability testing of triacs', *ibid.*, (5/6), pp. 254−9

2.1
PLEVYAK, T.J.: 'Vapour phase cooling of semiconductor circuit components', *IEEE Trans.*. 1969, IGA-5, pp. 607−612
GOLDEN, F.B.: 'Liquid cooling of power thyristors', IEEE Conf. 5th Annual Meeting, 1970, IGA, pp. 573−579
NEWELL, W.E.: 'Dissipation in solid state devices', *IEEE Trans.*, 1976, IA, (7/8), pp. 386−96
NEWELL, W.E.: 'Transient thermal analysis of solid state power devices', *IEEE Trans.*, 1976, IA, (7/8), pp. 405−20
MCLAUGHLIN, M.H., and VONZASTROW, E.E.: 'Power semiconductor equipment cooling methods and application criteria', *IEEE Trans.* 1975, IA, (9/10), pp. 546−55

2.2
ZAKEREVICIUS, R.A.: 'Calculation of the switching on overvoltages in a series connected thyristor string', *ibid.*, (9/10), pp. 407−17

2.4
LEACH, J.G., and NEWBERY, P.G.: 'Advances in development and application of semiconductor fuselinks', *Proc. IEE*, 1974, 123, pp. 1−6
SCHONHOLZER, E.T.: 'Fuse protection for power thyristors'. IEEE Conf. 5th Annual Meeting, 1970, IGA, pp. 455−465

3.2
CAVENDISH, L.F., and ZIELKE, R.A.: 'One year operating experience of large electrochemical SCR rectifiers', *ibid.* (3/4), pp. 147−50
ZEILKE, R.A.: 'A 50 MW thyristor controlled power converter', *IEEE Trans.,* 1975, IA (5/6), pp. 263−6

3.4
CREPAZ, S.: 'For an improved evaluation of conventional losses of transformers for convertors', *IEEE Trans.*, 1975, IA, (3/4), pp. 165−71

4.2
RAMSHAW, R.S., and PADIYAR, K.R.: 'Digital simulation of a full-wave single-phase converter system', *Proc. IEE*, 1970, 117, (11), pp. 2151−8

4.3
JONES, V.H., and BONWICK, M.E.: 'Three-phase bridge rectifiers with complex source impedance', *Proc. IEE*, 1975, 122, (6), pp. 630−6
WILLIAMS, S., and SMITH, I.R.: 'Fast digital computation of 3-phase thyristor bridge circuits', *Proc. IEE*, 1973, 120, (7), pp. 791−5
MCMURRAY, W.: 'A study of asymmetrical gating for phase controlled convertors', *IEEE Trans.*, 1972, IA, (5/6), pp. 289−95

4.4
REVANKAV, G.N., and MAHAJAN, S.A.: 'Digital simulation for mode identification in thyristor circuits', *Proc. IEE*, 1973, 120, (2), pp. 269−72

5.1
TURTON SMITH, B.: 'Thyristor control of traction battery chargers', *Proc. IEE,* 1974, 123, pp. 165−70
GAUPER, H.A., HARNDEN, J.D., and MCQUARRIE, A.M.: 'Power supply aspects of semiconductor equipment', *IEEE Spectrum*, 1971, 8, pp. 32−43
CULLIFORD, D.R.: 'Stabilised power supplies at Rutherford Laboratory',*Electronics & Power*, 1972, (4), pp. 133−5
BENNETT, P.A., and JONES, G.: 'Thyristors in action', *ibid.*, 1973, (9), pp. 399−402

MURPHY, J.D.: 'Solid state power rectification', *ibid.*, 1974, (10), pp. 898–901
MURPHY, J.D.: 'Semiconductor power', *ibid.* 1974, (2), pp. 57–60
CHOWDHURI, P., and WILLIAMSON, D.F.: 'Electrical interference for a thyristor controlled d.c. propulsion system of a transit car', *ibid.*, (11/12), pp. 539–50

5.2
SHEPHERD, W., and ZAKIKHANI, P.: 'Suggested definitions of reactive power for nonsinusoidal systems', *Proc. IEE*, 1972, 119, (9), pp. 1361–2
SHEPHERD, W., and ZAKIKHANI, P.: 'Reactive power definitions and power factor imporvement in nonlinear systems', *Proc. IEE*, 1974, 121, (5), pp. 390–2
NEDELAN, V.N.: 'Suggested definition of reactive power for nonsinusoidal systems', *Proc. IEE*, 1974, 121, (5), pp. 389–90
MICU, E.: 'Suggested definition of reactive power for nonsinusoidal systems', *Proc. IEE*, 1973, 120, (7), pp. 796–8
ZANDER, H.: 'Self-commutated rectifier to improve line conditions, *Proc. IEE*, 1973, 120, (9), pp. 977–81
SHARON, D.: 'Reactive power definitions and power factor improvement in nonlinear systems', *Proc. IEE*, 1973, 120, (6), pp. 704–5
SHARON, D.: 'Suggested definition of reactive power for nonsinusoidal systems', *Proc. IEE*, 1973, 120, (1), p. 108
SHEPHERD, W., and ZAKIKHANI, P.: 'Suggested definition for reactive power in nonsinusoidal systems', *Proc. IEE*, 1972, 119, (9), p. 1361
SHEPHERD, W., and KHALIL, A.Q: 'Capacitance compensation of thyristor-controlled slip-energy-recovery systems', *Proc. IEE*, 1970, 117, (5), pp. 948–56
MEHTA, P., MUKHOPADHYAY, S., and ORHUN , E.: 'Forced-commutated a.c.-d.c. converter-controlled d.c. drives', *Proc. IEE*, 1974, 123, pp. 146–152
ERIKSSON, L.G.: 'Thyristor control of multiple unit car equipment by ASEA'. IEEE Conf. 5th Annual Meeting, 1970, IGA, pp. 351–360
GALLOWAY, J.H.: 'Power factor and load characteristics for thyristor electro chemical rectifiers', *IEEE Trans.*, 1977, IA, (11/12), pp. 607–11
MOORE, A.H.: 'Application of power capacitors to electrochemical rectifier systems', *ibid*, (9/10), pp. 399–406

5.3
KRISHNAMURTHY, K.A., DUBEY, G.K., and REVANKOV, G.N.: 'Converter control with selective reduction of line harmonics', *Proc. IEE*, 1978, 125, (2), pp. 141–5
CORBYN, D.B.: 'The business of harmonics', *Electronics & Power*, 1972, (6), pp. 219-23
STEEPER, D.E., and STRATFORD, R.P.: 'Reactive compensation and harmonic suppression for industrial power systems using thyristor convertors', *IEEE Trans.*, 1976, IA, (5/6), pp. 232–54
ROHNSON, W.M.: 'Standards for commutating capacitors', *IEEE Trans.*, 1976, IA, (1/2), pp. 17–27
CHAMBERLAIN, T.: 'Harmonics and 3-phase thyristor drives', *Electronics & Power*, 1974, (10), pp. 838–41
MORI, H., SAWA, K., and IMURA, T.: 'Harmonic analysis of chopper controlled electric rolling stock', *IEEE Trans.*, 1973, IA, (5/6), pp. 302–9
GALLOWAY, J.H.: 'Live current waveforms and harmonics for a large multiphase thyristor converter systems', *ibid.*, (9/10), pp. 394–9
GALLOWAY, J.H.: 'Harmonic live currents in large thyristor 6-pulse convertors', *IEEE Trans.*, 1975, IA, (5/6), pp. 256–262

5.4
FARRER, W., and ANDREW, D.F.: 'Fully controlled regeneration bridges with half-controlled characteristics', *Proc. IEE*, , 1978, 125, (2), pp. 109–12
RAMAKUMAR, R., ALLISON, J.H., and HUGHES, W.L.: 'Analysis of the parallel bridge rectifier systems', *IEEE Trans.*, 1973, IA, (7/8), pp. 425–36
BEZOLD, K.H., FORSTER, J., and ZANDER, H.: 'Thyristor converters for traction d.c. motor drives', *IEEE Trans.*, 1973, IA, (9/10), pp. 612–7
MEHTA, P., and MUKHOPADHYAY, S.K.: 'Improvement in d.c. motor performance by asymmetrical triggering', *IEEE Trans.*, 1975, IA, (3/4), pp. 172–81

5.5
ALI, A.M.: 'Fast changing four-quadrant convertor', *Proc. IEE*, 1977, 124, (10), pp. 883–4

MEHTOR, P., and MUKHOPADHYAY, S.: 'Modes of operation in converter-controlled d.c. drives', *Proc. IEE*, 1974, 121, (3), pp. 179−83

HANCOCK, M.: 'Rectifier action with constant load voltages infinite capacitance condition', *Proc. IEE*, 1973, 120, (12), pp. 1529−30

CASTELL, R.J., and DANIELS, A.R.: 'Novel 4 quadrant thyristor controllers', *Proc. IEE*, 1972, 119, (11), pp. 1577−81

WILSON, J.W.A.: 'The forced commutated invertor as a regenerative rectifier', *IEEE Trans.*, 1978, IA, (7/8), pp. 335−40

5.6

MARTZLOFF, F.D., and HAHN, G.J.: 'Surge voltages in residential and industrial power circuits', *IEEE Trans.*, 1970, PAS-89, pp. 1049−55

MILLWARD, V.E., and RUSHALL, P.G.: 'Design and protection of AC/DC thyristor convertors', *Electronics & Power*, 1975 (6), pp. 728−32

CHOWDURI, P.: 'Transient voltage characteristics of silicon power rectifiers', *IEEE Trans.*, 1973, IA, (9/10), pp. 582−92

MCMURRAY, W.: 'Optimum snubbers for power semiconductors', *IEEE Trans.*, 1972, IA, (9/20), pp. 593−600

SCHONHOLZER, E.T.: 'Fuse protection for thyristors', *IEEE Trans.*, 1972, IA, (5/6), pp. 301-9

SILINGARDI, M.A.: 'Design criteria for the optimisation of series inductors in AC-DC thyristor convertors', *ibid.*, (1/2), pp. 141−7

LIJOU, A.L.: 'A megawatt converter with ride-through fault capability', *IEEE Trans.*, 1975, IA, (5/6), pp. 252−5

6.2

SHORE, N.L., and FRERIS, L.L.: 'Minicomputer on-line control of d.c. link converters', *Proc. IEE*, 1978, 125, (3), pp. 215−20

WILLIAMSON, A.C.,ISSA,N.A.H., and MAKKY, A.R.A.M.: 'Variable-speed inverter-fed synchronous motor employing natural conventation', *Proc. IEE*, 1978, 124, (3), pp. 213−7

WILLIAMSON, A.C., and CHALMERS, B.J.: 'New form of inverter-fed synchronous motors with induced excitation', *Proc. IEE*, 1977, 124, (3), pp. 213−7

CHALMERS, B.J., PACEY, K., and GIBSON, J.P.: 'Brushless d.c. traction drive', *Proc. IEE*, 1975, 122, (7), pp. 733−8

CHALMERS, B.J., MOHAMADEIN, A.L., and WILLIASON, A.C.: 'Inverter fed synchronous motors with induced excitation', *Proc. IEE*, 1974, 121, (12), pp. 1505−12

SCHOFIELD, J.R.G., and HALL, D.J.: 'Line communtated converters, protection from supply disturbances', *Proc. IEE*, 1974, 123, pp. 133−40

6.3

DANIELS, A.R., and SLATTERY, D.T.: 'Application of power transistors to poly-phase regenerative power converters', *Proc. IEE*, 1978, 125, (7), pp. 643−7

DANIELS, A.R., and SLATTERY, D.T.: 'New power converter technique employing power transistors', *Proc. IEE*, 1978, 125, (2), pp. 146−50

HANAD, A.K.S., HOLMES, P.G., and STEPHENS, R.G.: 'Phase-controlled circulating current cycloconvertor-induction-motor drive using rotating intergroup reactor', *Proc. IEE*, 1977, 124, (10), pp. 865−72

SMITH, G.A.: 'Static Scheibius system of induction motor speed control', *Proc. IEE*, 1977, 124, (6), pp. 557−60

RUBINAT, J.M., and ROCHET, A.: 'Thyristor naturally commutated converters for variable speed drives with high power a.c. machines', *Proc. IEE*, 1974, 123, pp. 160−64

DEWAN, S.B., and KANKAM, M.D.: 'A method for harmonic analysis of cyclo-converters', *IEEE Trans.*, 1970, IGA-6, pp. 455-462

TSUCHIYA, T.: 'Basic characteristics of cycloconverter-type commutatorless motors', *IEEE Trans.* 1970, IGA-6, pp. 349−356

PELLY, B.R.: 'Latest developments in statis high frequency power sources for induction heating', *IEEE Trans.* 1970, IECI-17, pp. 297−312

WURGLER, H.U.: 'The world's first gearless mill drive', *IEEE Trans.*, 1970, IGA-6, pp. 524−527

GYUGYI, L., and PELLY, B.R.: 'Static power frequency changers'

JACORIDES, L.J.: 'Analysis of a cycloconverter-induction motor drive system allowing for stator current discontinuities', *IEEE Trans.*, 1973, IA, (3/4), pp. 206−15

HAMBLIN, T.M., and BARTON, T.H.: 'Cycloconverter control circuits', *IEEE Trans.*, 1972, IA, (7/8), pp. 443–53

ESPELAGE, P.M., and BOSE, B.K.: 'High frequency link power conversion', *ibid.*, (9/10), pp. 387-94

CHATTOPADYAY, A.K.: 'An adjustable speed induction motor drive with a cyclo-convertor type thyristor commutator in the rotor', *IEEE Trans.*, 1978, IA, (3/4), pp. 116–22

6.4

ROMBAUT, C., SCHOOREUS, H., and SEGUIER, G.: 'Operation of a 3-phase a.c. thyristor regulator feeding an *R*, *RL* or *RC* balanced load', *Proc. IEE*, 1978, 125, (8), pp. 741–2

RAHMAN, S., and SHEPHERD, W.: 'Thyristor and diode controlled variable voltage drives for 3-phase induction motors', *Proc. IEE*, 1977, 124, (9), pp. 784–90

ZYBORSKI, J., CZUCHA, J., and SAJUACKI, M.: 'Thyristor circuit breaker for overcurrent protection of industrial d.c. power installations', *Proc. IEE*, 1976, 123, (7), pp. 685–9

LLOYD, S., and MARSHALL, P.: 'Dynamical performance of a single-phase thyristor a.c. regulator', *Proc. IEE*, 1975, 122,(12), pp. 1425–30

JACKSON, R.D., and WEATHERBY, R.D.: 'Thyristor alternating voltage controllers in closed-loop systems', *Proc. IEE*, 1974, 121, (9), pp. 1030–31

SHEPHERD, W., and GALLAGHER, P.J.: 'Power factor of thyristor-controlled single-phase resistive load', *Proc. IEE*, 1973, 120, (12), pp. 1538–9

SHEPHERD, W., and ZAKIKHANI, P.: 'Power factor compensation of thyristor controlled single-phase load', *Proc. IEE*, 1973, 120, (2), pp. 245–6

LLOYD, S.: 'Time series analysis of thyristor a.c. regulator', *Proc. IEE*, 1972, 119, (6), pp. 755–7

LINGARD, B.W., JOHNSON, R.W., and SHEPHERD, W.: 'Analysis of thyristor-controlled single-phase loads with integral cycle triggering', *Proc. IEE*, 1970, 117, (3), pp. 607–8

HARNDEN, J.D., and GOLDEN, F.B. (Eds.): 'Power semiconductor applications—Vol. 1 general considerations, Vol. 2 equipment and systems (IEEE Press, Selected Reprints, 1972)

RICH, D.A.: 'Thyristor control of a high voltage power supply for an electrostatic precipitator', *Proc. IEE*, 1974, 123, pp. 171–76

DAVIS, R.M., and DOWNING, B.R.: 'Integral cycle power control for 3-phase resistive loads', *Proc. IEE*, 1974, 123, pp. 177–82

MUSGRAVE, G., and O'KELLY, D.: 'Improvement of power system transmission by solid-state techniques', *Proc. IEE*, 1974, 123, pp. 228–233

LEE, T.H. SCHNEIDER, H.N., and TITUS, C.H.: 'Static switches for use in power systems and thermonuclear fusion research systems', *Proc. IEE*, 1974, 123, pp. 234–40

BEDFORD, R.E., and NENE, V.D.: 'Voltage control of the three-phase induction motor by thyristor switching: a time-domain analysis using the α-β-0 transformation', *IEEE Trans.*, 1970, IGA-6, pp. 553–562

BROADLEY, F.C., SHEPPARD, J.G., and SHEPHERD, W.: 'Solid state analysis of series *R-L* circuit controlled by asymmetrical triggering of thyristors', *IEEE Trans.*, 1973, IA, (7/8), pp. 437–47

PERRIN, E.M., and SCHÖNHOLZER, E.T.: 'Fundamental operation of rectifiers with thyristor AC power control', *IEEE Trans.*, 1973, IA, (7/8), pp. 453–61

MCMURRAY, W.: 'A comparative study of symmetrical 3-phase circuits for phase-controlled a.c. motor drives', *ibid.*, (5/6), pp. 403–11

SHEPHERD, W., and GALLAGHER, P.J.: 'Thyristor control of resistive and series d.c. motor loads using integral cycle switching', *ibid.*, (9/10), pp. 657–61

HIMEI, T., INOUYE, J., NAKANISHI, S., and UKITA, I.: 'A step-up phenomenon of the thyristor control circuit with series *RLC* elements', *IEEE Trans.*, 1975, IA, (9/10), pp. 531–39

KHALIFA, M., OBEID, M., and ENAMUL-HAQUE, S.: 'Effects of source impedance on the s.s. performance of thyristor controlled circuits', *IEEE Trans.*, 1975, IA, (7/8), pp. 384–91

7.3

SCUTCHINGS, J.H.: 'Wide range high performance time-ratio current control', *Proc. IEE*, 1977, 124, (8), pp. 709–714

DUCK, E.W., LAURENCE, L.S., and LOWE, T.J.: 'Thyristor chopper control and the introduction of harmonic currents into track circuits', *Proc. IEE*, 1976, 123, (6), pp. 523—30

FARRER, W.: 'D.C. to d.c. thyristor chopper for traction application', *Proc. IEE*, 1976, 123, (3), pp. 239—44

DUBEY, G.K., and SHEPHERD, W.: 'Analysis of d.c. series motor controlled by power pulses', *Proc. IEE*, 1975, 122, (12), pp. 1397—8

MCLELLAN, P.R.: 'Thyristor choppers using a bridge connected capacitor for commutation', *Proc. IEE*, 1975, 122, (5), pp. 514—6

LOWE, T.J., and MELLITT, B.: 'Thyristor chopper control and the introduction of harmonic current into track circuits', *Proc. IEE*, 1972, 121, (4), pp. 269—75

MELLITT, B., and RASHID, M.H.: 'Anlysis of d.c. chopper circuits by computer-based piecewise linear techniques', *Proc. IEE*, 1974, 121, (3), pp. 172—8

RICHARDSON, J.: 'New static controller for d.c. machines', *Proc. IEE*, 1972, 119, (11), pp. 1582—6

HALL, J.K.: 'A high-voltage d.c. chopper having individual commutation for the series-connected thyristors', *Proc. IEE*, 1974, 123, pp. 20—25

TSO, S.K., and AU, P.K.,: 'Analogue/hybrid simulation of a d.c. chopper drive system', *Proc. IEE*, 1974, 123, pp. 77—82

PARIMELALAGAN, R., and RAJAGOPALAN, V.: 'Steady state investigations of a chopper-fed d.c. motor with separate excitation', *IEEE Trans.*, 1971, IGA-7, pp. 101—108

REIMERS, E.: 'Design analysis of multi-phase d.c.-chopper motor drive', IEEE Conf. 5th Annual Meeting, 1970, IGA, pp. 587-595

FRANKLIN, P,W.: 'Theory of the d.c. motor controlled by power pulses, part 1 — motor operation'. IEEE Conf. 5th Annual Meeting, 1970, IGA, pp. 59—67

DALE-LACE. J.D.: 'Motor control by the use of thyristor chopper systems', *Electronics & Power*, 1974, (10), pp. 831—3

REIMERS, E.: 'An application of 2-phase d.c. chopper motor drive', *IEEE Trans.*, 1973, IA, (5/6), pp. 285—93

TSUBOI, T., IZAWA, S., WAJIMA, K., OGAWA, T., and KATTA, T.: 'Newly developed thyristor chopper equipment for electric railcars', *IEEE Trans.*, 1973, IA, (5/6), pp. 294—301

BERMAN, B.: '3 papers on battery vehicle choppers', *IEEE Trans.*, 1972, IA, (3/4), pp. 184—9, 190—4, 195—202

7.5

FARRER, W., and MISKIN, J.D.: 'Quasi-sine-wave fully-regenerative inverter', *Proc. IEE*, 1973, 120, (9), pp. 969—76

PEDDER, D.A.G.: 'Passive output filters for use with lightweight waveform synthesising inverters', *Proc. IEE*, 1974, 123, pp. 216—21

8.1

VERHOEF, A.: 'Basic forced commutated inverters and their characteristics', *IEEE Trans.*, 1973, IA, (9/10), pp. 601—6

CHAUPRADE, R.: 'Inverters for uninterruptible power supplies', *ibid.*, (7/8), pp. 281—97

8.2

VERHOEF, A.: 'Basic forced-commutated inverters and their characteristics', *IEEE Trans.*, 1973, IA (9/10), pp. 601—6

MCMURRAY, W.: 'A constant turn-off time control for variable frequency thyristor inverters', *ibid.*, (9/10), pp. 418—22

8.4

GUGGI, W.B.: 'Sine wave inverter system'. IEEE Conf., 1970, IGA, pp. 517—523

8.5

YAIR, A., and BEN URI, J.: 'New 3-phase inverter with 3 thyristors', *Proc. IEE*, 1971, 118, (7), pp. 901—5

BEDFORD, R.E., and NENE, V.D.: 'Analysis and performance of a three-phase ring inverter', *IEEE Trans.*, 1970, IGA-6, pp. 488—496

8.6

DAVIS, R.M., and MEELING, J.R.: 'Quantative comparison of commutation circuits for bridge inverters', *Proc. IEE*, 1977, 124, (3), pp. 237—246

265

PASHLEY, J.R.: 'Inverter techniques for power supplies and a.c. motor control, *Electronics & Power*, 1972, (6), pp. 206—9

FINNEY, D.: 'Static inverter systems for emergency supplies', *Electronics & Power*, 1973, (2), pp. 32—4

SULWAY, A.B.: 'Thyristor power supply equations for uninterruptible a.c. power', *Electronics & Power*, 1975, (11), pp. 1125—7

ABBONDANTI, A., and WOOD, P.: 'A criterion for performance comparison between high power inverter circuits', *ibid.*, (3/4), pp. 154—60

8.7

SRIRAGHAVEN, S.M., PRADHAM, B.D., and REVANKOV, G.N.: 'Multistage p.w.m. inverter system for generating stepped voltage waveforms', *Proc. IEE*, 1978, 125, (6), pp. 529—30

BALESTRINO, A., DE MARIA, G., and SCARVICCO, L.: 'Analysis of modulation processes in power convertors', *Proc. IEE*, 1978, 125, (5), pp. 411—2

BAUSAL, S.C., and RAO, V.M.: 'Evaluation of p.w.m. inverter schemes', *Proc. IEE*, 1978. 125, (4), pp. 328—334

LOCKWOOD, M.: 'Simulation of inverter/induction machine systems including discontinuous phase currents', *J. IEE Elec. Power & Appl.*, 1978, 1, (4), pp. 105—14

ALLIN, G., CREIGHTON, G.K., and HALL, J.K.: 'Operation and analysis of an inverter-fed linear motor system', *Proc. IEE*, 1972, 119, (11), pp. 1587—94

UDAYAGIRI, M.R., and SHEKHAWAT, S.S.: 'A novel stepped wave thyristor inverter', *Proc. IEE*, 1974, 123, pp. 152—60

ADAMS, R.D., and FOX, R.F.: 'Several modulation techniques PWM inverter'. IEEE Conf. 5th Annual Meeting, 1970, IGA, pp. 687—693

PENKOWSKI, L.J., and PRUZINSKY, K.E.: 'Fundamentals of a pulse width modulated power circuit'. IEEE Conf. 5th Annual Meeting, 1970, IGA, pp. 669—678

PATEL, H.S., and HOFT, R.G.: 'Generalised techniques of harmonic elimination and voltage control in thyristor inverters (Pt. 1)', *IEEE Trans.*, 1973, IA (S/C), pp. 310—7

LUNDU, N.D.: 'Basic operating characteristics of an h.f. inverter with capacitive voltage multiplier', *IEEE Trans.*, 1978, IA (5/6), pp. 264—8

MCMURRAY, W.: 'The performance of a single-phase current-fed inverter with counter-e.m.f. inductive load', *IEEE Trans.*, 1978, IA, (7/8), pp. 319—29

ZUBEK, J., ABBONDANTI, A., and NORBY, C.J.: 'PWM inverter motor drives with improved modulation', *IEEE Trans.*, 1975, IA (11/12), pp. 695—703

NAYAK, P.H., and HOFT, R.G.: 'Optimising the PWM waveform of a thyristor inverter', *IEEE Trans.*, 1975, IA, (9/10), pp. 526—30

LIPO, T.A., and TURNBULL, F.G.: 'Analysis and comparison of two types of square wave inverter drives', *IEEE Trans.*, 1975, IA, (3/4) pp. 137—47

8.9

EXCELL, P.S., and ATKINSON, P.,: 'Induction motor torques with p.w.m. inverter drive', *Proc. IEE*, 1977, 124, (7), pp. 646—7

GIBSON, J.P.: 'New inverter circuit suitable for high current p.w.m. operation', *Proc. IEE*, 1976, 123, (10), pp. 993—8

BOWES, S.R.: 'New sinusoidal p.w.m. inverter', *Proc. IEE*, 1975, 122, (11), pp. 1279—85

BOWES, S.R., and BIRD, B.M.: 'Novel approach to the analysis and synthesis of modulaton processes in power converters', *Proc. IEE*, 1975, 122, (5), pp. 507—13

EDWARDS, J.D.: 'Three phase digital p.w.m. inverter', *Proc. IEE*, 1975, 122, (3), pp. 302—4

BOWLER, P., and CHAN, T.Y.K.: 'An efficient high frequency inverter for a.c. drives', *Proc. IEE*, 1974, 123, pp. 127—132

ROBERTSON, S.D.T., and HEBBAR, K.M.: 'Torque pulsations in induction motors with inverter drives', *IEEE Trans.*, 1972, IGA ,7, pp. 318—323

BRENNEISEN, J., FUTTERLIEB, E., MULLER, E., and SCHULZ, M.: 'A new converter drive system for a diesel electric locomotive with asynchronous traction motors', *IEEE Trans.*, 1973, IA, (7/8), pp. 482—91

CARLTON, G.E., SKOGSHOLM, E.A., and VOLKMANN, W.K.: 'Practical design considerations for inverter drives', *IEEE Trans.*, 1973, IA, (9/10), pp. 593—600

THORBORG, K.: 'A 3-phase inverter with reactive power control', *IEEE Trans.*, 1973, IA, (7/8), pp. 473—81

PHILLIPS, K.P.: 'Current source converter for a.c. motor drives', *IEEE Trans.*, 1972, IA, (11/12), pp. 679—83

PENKOWSKI, L.J., and PRUZINSKY, K.E.: 'Fundamentals of a PWM power circuit', *IEEE Trans.*, 1972, IA, (9/10), pp. 584—92
POLLACK, J.J.: 'Advanced PWM inverter techniques', *IEEE Trans.*, 1972, IA, (3/4), pp. 145—54
PLONKETT, A.B.: 'Direct flux and torque regulation in a PWM inverter-induction motor drive', *IEEE Trans.*, 1977, IA, (3/4), pp. 139—46
SAWAKI, N., and NORIAKI, S.: 'Steady state and stability analysis of induction motor driven by current source inverter', *ibid.*, (5/6), pp. 244—53
HO, H.H.: 'A new and simple 3-phase thyristor inverter for variable frequency operation', *ibid.*, (11/12), pp. 787—96
PATEL, H.S., and HOLT, R.G.: 'Generalised techniques of harmonic elimination and voltage content in thyristor inverters', *ibid.*, (9/10), pp. 666—73

8.9
SLEMON, G.R., DEWAN, S.B., and WILSON, J.W.A.: 'Synchronous motor drive with current source inverter' *IEEE Trans.*, 1977, IA, (5/6), pp. 412—6

9.3.
TURNBULL, F.G.: 'A carrier frequency gating circuit for static inverters, converters and cycloconverters', *IEEE Trans.*, 1966, MAG-2, pp. 14—17

9.4
DANIELS, A.R., and LIPCZYNSKI,R.T.: 'Digital firing angle circuit for thyristor motor controllers', *Proc. IEE*, 1978, 125, (3), pp. 245—6
SAHM, W.H.: 'A new triac trigger to optimise consumer phase controls', *IEEE Trans.*, 1973, IA, (1/2), pp. 46—8

9.5
DINGER, E.H.: 'Textile applications of adjustable frequency drives with digital ratio control', *IEEE Trans.*, 1972, IA, (1/2), pp. 47—55

9.7
SKJELLNES, A., HANSSEN, B., and ARNULF, T.: 'Phase locked loop control of thyristor convertors', *Proc. IEE*, 1976, 123, (10), pp. 999—1001
DE, G., and MANDAL, A.K.: 'Impulse analysis of subharmonic oscillations in control systems with thyristor convertors', *Proc. IEE*, 1973, 120, (9), pp. 1030—7
SUCENA-PAIVA, J.P., and FRERIS, L.L: 'Stability of rectifiers with voltage-controlled oscillator firing systems', *Proc. IEE*, 1973, 120, (6), pp. 659—66
SUCENA-PAIVA, J.P., HERNANDEZ, R., and FRERIS, L.L.: 'Stability study of controlled rectifiers using a new discrete model', *Proc. IEE*, 1972, 119, (9), pp. 1285—93
FLOWER, J.O., and HAZELL, P.A.: 'Nonlinear analysis of a first order thyristor bridge control system', *Proc. IEE*, 1971, 118, (10), pp. 1511—6
CHEUNG, W.N.N.: 'Realisation of convertor control using sample and delay method', *Proc. IEE*, 1971, 118, (5), pp. 701—5
HAZELL, P.A., and FLOWER, J.O.: 'Theoretical analysis of harmonic instability in a.c.-d.c. converters', *Proc. IEE*m 1970, 117, (9), pp. 1869—78
AINSWORTH, J.D.: 'The phase-locked oscillator—a new control system for controlled static convertors', *IEEE Trans.*, 1968, PAS-87, pp. 859—65
LIPO, T.A., and CORNELL, E.P.: 'State variable s.s. analysis of a controlled current induction motor drive', *IEEE Trans.*, 1975, IA, (11/12), pp. 704—12
General
NEWELL, W.C.: 'Power—emerging from limbo', *IEEE Trans*, 1975, IA, (1/2), pp. 7—11

INDEX